MAGNETIC MEASUREMENTS ON ORGANIC COMPOUNDS

N.V. VAN DE GARDE & CO'S DRUKKERIJ, ZALTBOMMEL

MAGNETIC MEASUREMENTS ON ORGANIC COMPOUNDS

PROEFSCHRIFT

TER VERKRIJGING VAN DE GRAAD VAN
DOCTOR IN DE TECHNISCHE WETENSCHAP
AAN DE TECHNISCHE HOOGESCHOOL TE
DELFT KRACHTENS ARTIKEL 2 VAN HET KO-
NINKLIJK BESLUIT VAN 16 SEPTEMBER 1927,
STAATSBLAD NO. 310 EN OP GEZAG VAN DEN
RECTOR MAGNIFICUS J. M. TIENSTRA, HOOG-
LEERAAR IN DE AFDELING DER WEG- EN
WATERBOUWKUNDE VOOR EEN COMMISSIE
UIT DE SENAAT TE VERDEDIGEN OP DON-
DERDAG 9 JANUARI 1947, DES NAMIDDAGS
TE 4 UUR

DOOR

SIJBRAND BROERSMA

GEBOREN TE HARLINGEN

SPRINGER-SCIENCE+BUSINESS MEDIA, B.V.
1947

Dit proefschrift en de stellingen zijn goedgekeurd door den promotor
Prof. Dr. R. Kronig

ISBN 978-94-017-6713-2   ISBN 978-94-017-6790-3 (eBook)
DOI 10.1007/978-94-017-6790-3

CONTENTS

CONTENTS

OBJECT OF THE INVESTIGATION

In the work described in this thesis we had the object of performing precise and swift measurements of magnetic susceptibilities; the accuracy aimed at being $0.7 \cdot 10^{-8}$, one pro mille of the diamagnetic susceptibility $(\varkappa)$. After a brief discussion of the theory (I) [1] and a survey of the experimental methods hitherto employed, including those of the author (II), the major part of our considerations is devoted to the description of apparatus, among which we mention a torsion balance (III) as well as a standard and a new type of inductance apparatus (IV and V). These methods were employed in the measurement of a great number of organic compounds. A discussion of the calculation and errors of our measurements is given (VI) as well as the results for pure compounds (VII) and mixtures (VIII).

We controlled the results of other authors, in many respects still discrepant, and for this reason applied two methods of measurement to the investigation of diamagnetic substances.

A further application of our investigation may be found in two other directions, viz. the employment of the law of C u r i e in paramagnetic substances for exact temperature determination at high temperatures, which formed part of a research program of the late Prof. dr. E. C. W i e r s m a (1936) (see V.3). Also as suggested by Prof. dr. R. K r o n i g, the statistical analysis of hydrocarbon mixtures, developed by Prof. dr. ir. H. I. W a t e r m a n (1935), might be supplemented by data on the magnetic susceptibility (see VIII.3).

REFERENCES

W a t e r m a n, H. I., c.s. (1935). J. Inst. Petr. Techn. **21**, 661 and 701; **24**, 16.
W i e r s m a, E. C. (1936). Eenige punten uit de ontwikkeling van het temperatuur-
begrip.

---

[1] In this thesis Roman ciphers indicate the chapters, Arabian ciphers the paragraphs of each chapter. Equations and figures are numbered anew in each chapter.

CHAPTER I

THEORY

1. Introduction

We start with the equations of Maxwell, expressed in the rationalised Giorgi system [1]:

$$\text{curl } \mathbf{H} = \mathbf{I} + \frac{\partial \mathbf{D}}{\partial t}, \quad (1) \qquad \text{curl } \mathbf{E} = -\frac{\partial \mathbf{B}}{\partial t}, \quad (2)$$

$$\text{div } \mathbf{D} = Q, \quad (3) \qquad \text{div } \mathbf{B} = 0. \quad (4)$$

The magnetic field is described by means of the two field quantities $\mathbf{B}$ and $\mathbf{H}$. $\mathbf{B}$ is expressed in Vsec/m^2 and $\mathbf{H}$ in A/m. Between them the following relation exists:

$$\mathbf{B} = \mu_0 \mu_r \mathbf{H} = \mu_0 \mathbf{H} + \mathbf{J}, \quad (5)$$

where $\mu_0 - 4\pi \cdot 10^{-7}$. $\mathbf{J}$, the intensity of magnetisation [2] in the medium, often is small so that we can better examine the relation between $\mathbf{J}$ and $\mathbf{H}$, according to (5):

$$\mathbf{J} = \mu_0(\mu_r - 1) \mathbf{H} = \varkappa \mu_0 \mathbf{H}. \quad (6)$$

$\varkappa$ is the *magnetic susceptibility* of the medium, referred to a unit volume. Dividing by the density ρ, we obtain the susceptibility referred to unit mass:

$$\chi = \frac{\varkappa}{\rho}. \quad (7)$$

In a uniformly magnetised sample the magnetic moment is defined by

$$\mathbf{M} = v\mathbf{J} = v\varkappa\mu_0 \mathbf{H} = m\chi\mu_0 \mathbf{H} = n\chi_M\mu_0 \mathbf{H}, \quad (8)$$

---

[1] In this system the meter (m), kilogram (kg), second (sec; 1sec^{-1} = 1 Hz), coulomb (C), ampère (A), volt (V), ohm (Ω), henry (H), farad (F) are used. For comparison: B (expressed in Vsec/m^2) = $10^{-4} \cdot$ B (expr. in gauss); H (expr. in A/m) = $10^3/4\pi \cdot$ H (expr. in gauss).

[2] In the literature other definitions are sometimes given, but the close connection of $\mathbf{J}$ with $\mathbf{B}$ seems to be the most convenient.

in which v is the volume, m the mass and n the number of kilomols of the sample. χ_M is the susceptibility per kilomol. As a measure we give the value for water:

$$\varkappa(20^0) = -903 \cdot 10^{-8}, \; \chi = -0.905 \cdot 10^{-8}, \; \chi_M = 16.30 \cdot 10^{-8}.$$

Organic compounds usually have a somewhat smaller magnetic susceptibility. Below we shall always use -10^{-8} as unit for the susceptibility.

2. Formula of V a n Vl e c k

Deriving the equations of M a x w e l l as macroscopic field equations from Lorentz' equations for the microscopic field, it appears that (5) and (8) correspond to the following definition of the magnetic moment of a system of moving charges:

$$\mathbf{m} = \underset{i}{\Sigma} \tfrac{1}{2} \mu_0 \left[\mathbf{r}_i, \, e_i \, \mathbf{v}_i \right], \tag{9}$$

where $\mathbf{v}_i$ is the velocity and $\mathbf{r}_i$ the radiusvector of the i-th charge e_i. Such moving charges are present in atoms and the magnetic properties of the substances arise from them.

As is well-known, a theory of para- and diamagnetism on the basis of classical physics was first given by L a n g e v i n. With the arrival of the quantum theory it became necessary to translate his results into the new language. V a n V l e c k (1932) was the first who gave a rigorous calculation of the susceptibility of an assembly of independent gas molecules or ions in solutions.

His starting point is the classical H a m i l t o n function of moving charges in magnetic and electric fields, which in quantum mechanics must be considered as a matrix. The magnetic moment also becomes a matrix, the diagonal elements of which represent the magnetic moments of the atoms in their stationary states while the non-diagonal elements are related to transitions to other states. According to v a n V l e c k these elements can be divided into two groups, viz. the low and high frequency elements, depending on whether the energy difference between the states is small or large compared with kT. In general the atoms, molecules or ions can be supposed to be in their lowest electronic states.

The low frequency elements give an effect, depending on the temperature, of the same kind as is given by Langevin's theory. In the mathematical expression of this effect they are represented by the permanent magnetic moment present in the atoms. The high frequency elements give an effect independent of the temperature. They can be characterised by the quantum numbers, collectively denoted by n and n', of the two states involved.

The formula obtained by v a n V l e c k is:

$$\chi_M = \frac{Nm^2}{3\mu_0 kT} + \frac{2N}{3\mu_0} \Sigma' \frac{|\mathbf{m}(n, n')|^2}{h\nu(n', n)} - \mu_0 \frac{N}{6} \Sigma_i \frac{e_i^2}{m_i} \overline{r_i^2} =$$

$$\chi_M^{(1)} + \chi_M^{(2)} + \chi_M^{(3)}. \qquad (10)$$

Here N is the number of A v o g a d r o, e_i and m_i the charge and mass of the i-th electron or nucleus, h and k the constants of P l a n c k and B o l t z m a n n, $\mathbf{m}$ the permanent magnetic moment occurring in the atoms, $\mathbf{m}(n, n')$ the high frequency elements and $h\nu(n', n)$ the energy difference of the two states between which the transition occurs ($n = n'$ is excluded). The effect of the nucleus in $\chi_M^{(3)}$ is small.

Notwithstanding the fact that several simplifications have been introduced, (10) still gives a general view of the existing effects. Apart from the paramagnetic terms, $\chi_M^{(1)}$ depending on and $\chi_M^{(2)}$ independent of the temperature, the negative diamagnetic term $\chi_M^{(3)}$ is present.

V a n V l e c k (1932, page 276) proved that the sum of the diamagnetic and high frequency term is invariant of the choice of the zeropoint of r. If we have independent atoms, we can take their nucleï as zero-point. It is proved that the high frequency term then is zero. Forming molecules from independent atoms, we can try to describe the new situation by using for each atom its own nucleus as zero-point and correcting the diamagnetic term for changes in the charge distribution, at the same time adding a high frequency term, arising from the non-central fields due to the adjoining atoms.

If these extra terms are smaller than the main effect, it is obvious that an additivity rule is fulfilled in a first approximation. This is found experimentally.

Furthermore the attractive forces between the atoms in general will reduce their effective volume, both in the formation of molecules and on condensing the gasmolecules to liquids. As is well-known, the transition to the solid state again is accompanied by an increase in the density so that on passing from free atoms to atoms bound in the solid state, a consecutive decrease of r^2 is rather plausible. On the other hand the contribution connected with the high frequency elements, starting from zero, increases in this series due to the increase of the mutual perturbation of the atoms and molecules. Therefore as a rule we can expect a decrease of $|\chi_M^{(2)} + \chi_M^{(3)}|$ on forming molecules from atoms and condensing them to solids. In special cases paramagnetism may even arise. If deviations from this general behaviour occur, they often are of great interest, e.g. in the formation of aromatic molecules from carbon atoms or when association occurs in liquids.

3. The susceptibility of assemblies of independent atoms

If there are resultant orbital and spin moments, the first term in (10) is large compared with the last term; the second one disappears in atoms. **m** is conveniently expressed in the B o h r magneton $\mathbf{m}_B = \mu_0 \, e/2m \cdot h/2\pi$, the magnetic moment of a p-electron in its orbit. Then the term dependent on the temperature becomes:

$$\chi_M^{(1)} = \frac{N^2 g^2 J(J+1)}{3\mu_0 \, RT} \, \mathbf{m}_B^2, \tag{11}$$

in which g is the L a n d é factor and J the quantum number defining the total angular momentum in the atom. In the classical theory of L a n g e v i n a term of this type is also found as the result of the directional effect of the magnetic field upon the permanent magnetic moments, impeded by the heat motion. Therefore the temperature appears in this term. At room temperature χ may be a factor 10 to 100 larger than the diamagnetic susceptibility and of course with opposite sign.

The *diamagnetic* term is always present. When the magnetic field is put on, an e.m.f. will be generated according to (2). From this the L a r m o r precession results, giving rise to a magnetic moment opposite to the original field so that the substance is repelled by the magnet, provided other effects are absent. As the induction effect is not influenced by the heat motion, the diamagnetic susceptibility is independent of temperature. The range of susceptibilities is rather small.

The diamagnetic effect can be calculated exactly for hydrogen-like atoms (v a n V l e c k and P a u l i n g). For atomic hydrogen in its normal state they find

$$\chi_H = -\mu_0 \frac{Ne^2}{2m} a_0^2 = -2.98 \cdot 10^{-8}, \tag{12}$$

where $a_0 = \mu_0 h^2 / \pi m e^2 = 52.9$ pm, is the radius of the smallest orbit in B o h r's original theory of the H-atom.

As is obvious, χ_H will be a convenient measure for the diamagnetic term in (10), which we can write as

$$\chi_M^{(3)} = \tfrac{1}{3} \chi_H \sum_i \overline{\left(\frac{r_i}{a_0}\right)^2}. \tag{13}$$

Several calculations were performed on other atoms (see e.g. S t o n e r, 1934). In table I we give a short survey of the calculations for some inert gases. Even for these simple cases the calculated values differ rather much from the experimental ones.

TABLE I

	He	A	X
P a u l i n g (1927).	1.94		31.9
H a r t r e e-S t o n e r (1929) . .	2.39	31.2	
S l a t e r (1930)	2.06	23.2	60.3
A n g u s (1932)	2.11	21.0	56.3
K i r k w o o d-V i n t i (1932). .	2.48	21.0	57.2
measured	2.39	24.5	53.3

TABLE II

$(1s)^2$	0.0
$(2s)^2 (2p)^6$	0.2
$(3s)^2 (3p)^6$	1.7
$(3d)^{10}$	6.7
$(4s)^2$	6.3
$(4p)^6$	31.2
Br^-	46.1

In table II the results of the calculations of A n g u s for Br^- are given, showing the contributions of the various types of electrons. As can be seen, the outer electrons give the main contribution, $\tfrac{1}{4}$ of the charge causing $\tfrac{2}{3}$ of the effect. Besides the fact that for larger values of the quantum number the size of the orbits increases, the effective nuclear charge is the smallest for the outer electrons, causing an additional increment in the radius of the orbit.

4. The susceptibility of assemblies of independent molecules

In organic molecules the *paramagnetic term* occurs for free radicals and biradicals. In the former case one, in the latter two isolated free valencies exist, corresponding to an equal number of electrons with spins. Usually the molecules carry no orbital moments so that the spin-only value($S = \tfrac{1}{2}$) becomes:

$$\chi_M = \frac{N^2 \, m_B^2}{\mu_0 R T}. \tag{14}$$

For biradicals twice this value should be found (see M ü l l e r, 1935).

In analogy with dielectric effects the *high frequency term* of (10) is often called the magnetic polarisation. Its effect increases as the non-central influence of the nuclear structure becomes more pronounced. As it occurs always in molecules, it will be of interest to express this term also in the susceptibility of hydrogen atoms.

In order to estimate its magnitude we shall replace $v(n', n)$ by an average frequency v_0, for which we write with the aid of R y d-b e r g's constant $R = h/8\pi^2 mca_0^2$:

$$v_0 = \alpha Rc,$$

v_0 being in general of the order of the resonance frequency of hydrogen and hence α of the order 1. Now we can put

$$\underset{n'(n' \neq n)}{\Sigma} |\, \mathbf{m}(n, n')\,|^2 = \mathbf{m}^2(n, n) = \left(g\, \mu_0 \frac{e}{2m} \cdot \frac{h}{2\pi}\right)^2 P^2\,(n, n),$$

where $P(n, n)$ is the angular momentum, measured in units $h/2\pi$. We then find on combining it with the *diamagnetic term*:

$$\chi_M^{(2)} + \chi_M^{(3)} = \tfrac{1}{3}\chi_H \left[\Sigma \overline{\left(\frac{r_i}{a_0}\right)^2} - 2\,\frac{g^2 P^2(n, n)}{\alpha}\right]. \qquad (15)$$

We shall consider some special cases of diamagnetic molecules. The effect for *hydrogen molecules* has been calculated with (15). Refractive data yield $\alpha = 1.23$; while $g = 1$. V a n V l e c k and F r a n k (1929) applied known wave functions for the hydrogen m o l e c u l e s (W a n g, 1928). Their theoretical values for the susceptibility are given in table III.

TABLE III

Unit —10^{-8}	$\chi^{(3)}$	$\chi^{(2)}$	$\chi^{(3)} + \chi^{(2)}$
V a n Vl e c k (1929) .	5.92	—0.64	5.28
W i t m e r (1942) . . .	5.17	—0.12	5.05
experimental			5.03
two H atoms	5.96		5.96

The value of W i t m e r (1942) agrees rather well with the experimental results. We see that the effect for the molecules is 0.85 of the total effect of two isolated atoms.

In this molecule the high frequency part is only a small portion of the effect. A similar situation prevails also for the F a r a d a y effect (see S c h ü t z). It appears there that B e c q u e r e l's formula describes the experiments on hydrogen within a precision of 10 pro mille. Other compounds show an anomalous factor of as much as 0.5, suggesting that the effect of the high frequency terms is much larger than in hydrogen.

A very interesting case is the *aromatic ring*. K r i s h n a n and collaborators (1936), observing crystals, found that the susceptibility in a direction perpendicular to the planes of the aromatic molecules was very large, in the planes themselves normal, e.g. in the case of benzene 118 and 43 respectively. L o n d o n (1937), after a fairly rigorous calculation, obtained for the effect of the ring of benzene, in a direction perpendicular to the plane of the ring:

$$\chi_{ring} = -\,\mu_0\,NW\left(\frac{\pi\,ear}{h}\right)^2,$$

where W is a sort of interaction energy of neighbouring atoms. Here a is the distance of two neighbouring carbon atoms in the ring, r the radius of the ring. We shall compare a with the radius of hydrogen and hence write:

$$a = 2\beta\,a_0.$$

Also we express W in terms of the energy of hydrogen in its normal state:

$$W = \gamma\,\frac{me^4}{8\mu_0^2\,h^2}.$$

Then it follows that

$$\chi_{ring} = \beta^2\gamma\chi_H\left(\frac{r}{a_0}\right)^2 = 2\beta^2\gamma\cdot\tfrac{1}{2}\chi_H\left(\frac{r}{a_0}\right)^2. \qquad (16)$$

This formula agrees rather well with (13). A factor 3/2 has to be introduced, as here all aromatic rings are perpendicular to the magnetic field. The magnetic effect, due to the ring, placed perpendicular to the field, is 75; $\chi_H = 2.98$ while X-ray investigations show that the radius of the aromatic ring is 139 pm. From this there follows $2\beta^2\gamma = 7.3$ (as the radius of an aromatic C-atom is 69.5 pm, γ is about 2). Applying (13) this would be the effect of 7.3 electrons while 6, one from each carbon atom, are present. In each direction the other 3 carbon electrons give a contribution to the susceptibility.

5. The susceptibility of the liquid and solid state

The previous discussion refers to separate molecules. A theoretical treatment of the susceptibility in the liquid state, on account of the interactions of the molecules, meets still greater difficulties so that guidance and control by experimental facts cannot be done without. We discuss the rather limited evidence first.

Some authors compared the liquid and the gaseous phase, obtaining indications of an increase in the susceptibility when going to the gaseous state.

The influence of the interaction can be investigated further by examining *the dependence of the susceptibility on temperature* and *on transitions between the liquid and solid state* as well as by *diluting liquids with other liquids*. We shall describe several typical features that were observed. It must be noticed, however, that the experimental material is not very extensive and that the results of different authors do not agree within their precision of measurement.

Pure liquids

a. C a b r e r a and F a h l e n b r a c h (1933) found for the dependence on the temperature of the susceptibility of several aliphatic alcohols the curve sketched in fig. 1. Far below the melting-

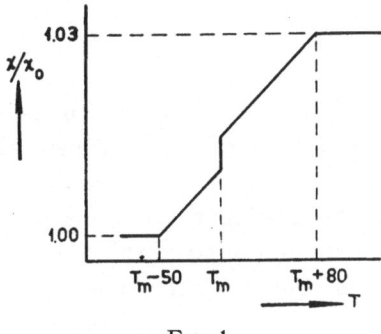

Fig. 1

point (T_m) χ is constant as well as far above. In between increases about 30 pro mille ($+0.2$ pro mille per °C), while the jump at T_m amounts to 0.2 of the total increase. For octyl alcohol ($C_8H_{17}OH$) and dodecyl alohol (C_{12}) they found the whole curve. For methyl alcohol (C_1) they checked the constancy of the right-hand part far above T_m, for cetyl alcohol (C_{16}) only the left-hand part.

b. A second group is formed by the aromatic compounds. They show a dependence on temperature as is drawn in fig. 2. The effects are much larger. The jump at the melting-point can amount to 200 pro mille, while there is a correlation with the size of the electric dipole moment. The dotted line indicates the possibility of under-cooling. R a o and S r i r a m a n (1933) found for nitrobenzene for temperatures above the melting-point the right-hand part of the curve. They ascribe the effects to association occurring in the liquid. If the temperature rises, the association effects are suppressed and the susceptibility again increases.

The values mentioned in literature for the dependence on the temperature for this group of liquids range from —1 to +1 pro mille per °C (B h a t n a g a r, 1934; A z i m, 1933).

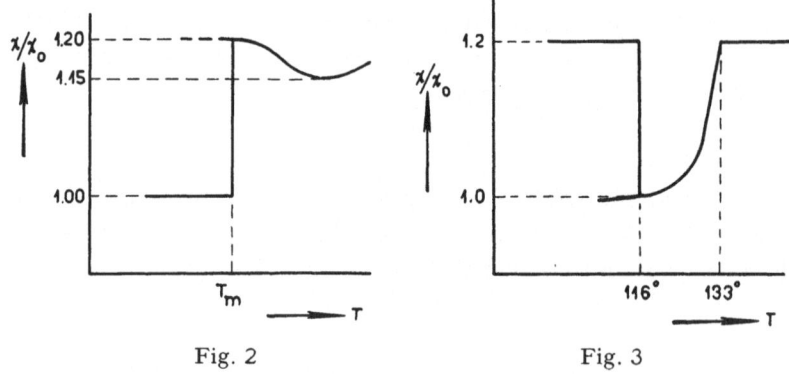

Fig. 2   Fig. 3

A typical substance, showing the effects of the anisotropy of the aromatic compounds, is para azoxy-anisol, a so-called liquid crystal (fig. 3). Just above the melting-point the solid complexes direct themselves so that the diamagnetism decreases. With increasing temperature the heat motion opposes this and the original value is approached again. On cooling to the solid state the crystals will remain directed.

c. A third group consists of compounds, containing neither electric dipoles nor aromatic rings, e.g. benzene, hexane and carbon tetra-chloride. The dependence on the temperature lies between —0.05 and +0.05 pro mille per °C. The change on melting is small.

Mixtures

When there are no dipoles, no association effects occur and on mixing the liquids the susceptibility will be linearly related with the concentration. The susceptibility of the mixtures will not depend upon the temperature. When there are dipoles, the non-linearities appearing on mixing may amount up to 10 pro mille while the dependence on the temperature may be 0.2 pro mille per °C.

F a r q u h a r s o n (1931) measured aequeous solutions of acids. Diluting e.g. HCl, he found extreme values in the susceptibility for the molecular concentrations acid: water = 1 : 1, 1 : 2 up to 1 : 10. For the ratio 1 Cl-ion to 6 watermolecules the effect was fairly pronounced.

Results

As a result we can recapitulate that, apart from some interaction effects which often are small, the susceptibilty of liquids and solids coincides fairly well with the properties of the single molecules.

In 2, in the discussion of the equation of v a n V l e c k, it appeared that the susceptibility of an assembly of molecules can be considered as consisting of a contribution of free atoms or ions and a part caused by the mutual perturbation of the atoms in the molecules. This will be specially true in organic chemistry with its large number of molecules, consisting of the same atoms and having a similar structure. Especially the homologous series are of interest in this respect. On the other hand it might be useful for the application not to use free atoms as basis, but the contribution of the atoms or even groups in the form in which they most often occur in the molecules; thus e.g. the CH_2-group in a long chain could be used as basis. The magnetic effect of the perturbation, e.g. on forming rings or double bonds, could then be taken together and considered as giving its own contribution to the susceptibility. Hence on starting with the equation

$$\varkappa(t) \frac{M}{\rho(t)} = \chi_M = \sum_k n_k \chi_M^k,$$

where n_k is the number of contributing elements of the kind k present in the molecule, χ_M^k then represents the contribution to the susceptibility of a certain atom or group of atoms, occurring in the molecules, but also the contribution due to a constitutive element. Considering mixtures, χ_M^k represents the effect of a pure compound, n_k the relative content of that compound.

For much experimental work on this subject the extensive investigations of P a s c a l (1905–1915) form the basis (see e.g. K l e m m, 1936). He found that a relatively small number of values χ_M^k suffice to express χ_M within his precision of measurement, being perhaps 30 pro mille. Especially homologous series could well be described in this way. The present day measurements, having an accuracy of perhaps 3 pro mille, often cannot be interpreted with a similiar precision as the typical features of the molecules and their inter-action then become noticeable.

We have observed the homologous saturated hydrocarbons and the effect of rings and ramifications, furthermore some homologous acids, alcohols and other compounds. The effect of stereo-isomerism is examined in some sugars. We have also investigated the effects of mixing and tried an application to statistical analysis.

REFERENCES

A n g u s, R. (1932). Proc. Roy. Soc. (A) **136**, 569.
A z i m, M. A., S. S. B h a t n a g a r and R. N. M a t h u r (1933). Phil. Mag. (7) **16**, 580.
B h a t n a g a r, S. S., M. B. N e v g i and M. L. K h a r n a (1934). Z. Phys. **89**, 506.
C a b r e r a, B. and H. F a h l e n b r a c h (1933). Z. Phys. **85**, 568 and **89**, 682.
F a r q u h a r s o n, J. (1931). Phil. Mag. (7) **12**, 283.
H a r t r e e, D. R. (1928). Proc. Camb. Phil. Soc. **24**, 89 and 111.
K i r k w o o d, J. G. (1932). Phys. Z. **33**, 57.
K l e m m, W. (1936). Magnetochemie.
K r i s h n a n, K. S. and K. L o n s d a l e (1936). Proc. Roy. Soc. (A) **156**, 597.
L o n d o n, F. (1937). C. R. **205**, 28.
M ü l l e r, E. and I. M ü l l e r- R o d l o f f (1935). Lieb. Ann. **517**, 134.
P a u l i n g, L. (1927). Proc. Roy. Soc. (A) **114**, 181.
R a o, S. R. and S. S r i r a m a n (1933). Ind. Jour. Phys. **8**, 315.
S c h ü t z, W. Magneto-optik. (Hb. der Exp. phys. XVI. 1).
S l a t e r, J. C. (1930). Phys. Rev. **36**, 57.
S t o n e r, E. C. (1929). Proc. Leeds Phil. Soc. **1**, 484.
S t o n e r, E. C. (1934). Magnetism and matter.
V i n t i, J. P. (1932). Phys. Rev. **41**, 813.
V l e c k, J. H. v a n and H. F r a n k (1929). Proc. Nat. Acad. **15**, 539.
V l e c k, J. H. v a n (1932). The theory of electric and magnetic susceptibilities.
W a n g, S. C. (1928). Phys. Rev. **31**, 579.
W i t m e r, E. E. (1942). Phys. Rev. **61**, 387.

CHAPTER II

SURVEY OF THE EXPERIMENTAL METHODS

1. General remarks

Two methods are applied in the measurement of susceptibilities. In the first place the *ponderomotive force*, exerted by a magnet upon a substance brought into the magnetic field, can be used. If the substance has a degree of freedom in the direction of the force, it will be moved, unless a second force opposes this motion so that the body stays in its original position. According to the second method one measures the *electromotive force*, generated in an alternating magnetic field. The e.m.f. changes if a magnetisable body is brought into the field. The currents arising from this e.m.f. in a conductor can be observed with a galvanometer and compensated by applying another e.m.f. It is of importance for the measurement that the ponderomotive force or the e.m.f. by which the effect is compensated should be continuously variable and linearly related with another quantity, such as a resistance, that can directly be read off. Instead of compensation the measurement of the frequency of oscillation of the system is often applied.

In connection with the fact that usually several compounds have to be compared it is desirable that an exchange of the substance can easily be accomplished. It is clear that only the relative position of the magnetic field and the substance determines the effect. Therefore in the force measurements there are two possibilities, a movable sample or a movable magnet. In the induction method such a choice also occurs as will be seen below. At first sight the induction method, as a field measurement, would seem to be somewhat more suited for a swift replacing of the sample.

2. Force methods

The force on a magnetised body, placed in an inhomogeneous

magnetic field ($\mathbf{H}_0$), is given by

$$\mathbf{F} = (\mathbf{M} \cdot \text{grad})\, \mathbf{H}_0, \tag{1}$$

provided the dimensions of the body are sufficiently small. On this equation all force measurements are based.

Hereafter we shall mention some of the methods applied in the force measurements. The systems used differ in three respects: the way the movable part is suspended, the manner in which the magnetic field is applied and the form in which the measurement is carried out, such as e.g. compensation. A discussion will be given under the headings *mechanical part, magnetic part, measurement.*

Mechanical part

As is obvious, the number of degrees of freedom of the system has to be reduced to one, while keeping the frictional force in the remaining direction as small as possible. Now outer influences can only give displacements in that direction. Disturbing effects may be due to vibrations of the base of the apparatus or introduced by switching of the current before equilibrium of the forces has been reached. As we are working in constant fields, we can decrease the oscillations by adding a dissipative agency, such as air damping or magnetic damping; the latter agency is used by us, as it can easily be varied.

In the following we briefly discuss some systems.

a. W i e r s m a (1930) suspended a small ball on a wire (fig. 1).

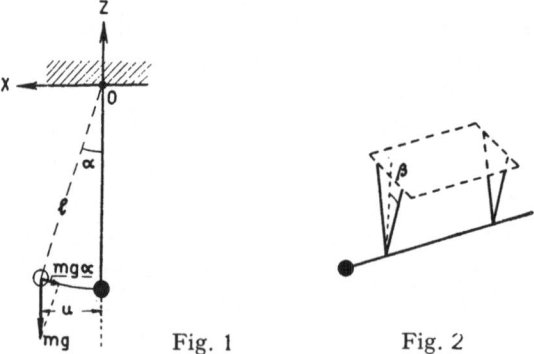

Fig. 1   Fig. 2

If F is the force, keeping the ball at its place, then

$$- F_x = mg \sin \alpha \approx mg\,\alpha = \frac{mg}{l} u = \frac{2 \cdot 10^{-2}}{0.3} u = 6 \cdot 10^{-2} u \text{ Newton,} \tag{2}$$

in which u is the displacement that has to be observed. We assumed a mass of 2 g. As the magnetic force is also proportional to m, in this method u is independent of the quantity of the sample.

The system is only sensitive for a displacement in a horizontal direction, that can be compensated by an opposite one of the point 0. As the disturbing influences in the y-direction may be large, W e i s s and F o ë x (1911) took the angle α in this direction (β in fig. 2) different from zero. They suspended a bar on four wires and so only displacements in one horizontal direction were possible (see also W i e r s m a and W o l t j e r, 1929).

b. The t o r s i o n b a l a n c e (fig. 3) was used by C u r i e and C h é n e v e a u (1903). The measuring system is hanging on a vertical torsional wire while at the lower side also a torsional wire can be attached. The wire can be taken thick enough to stand the weight of the system, but nevertheless gives only a small torsional

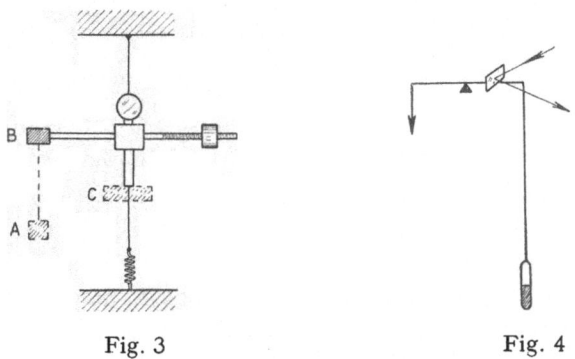

Fig. 3   Fig. 4

constant. If we assume e.g. $D = 1 \cdot 10^{-4}$ Nm/rad (in III D could still be chosen 10 times smaller) and use a lever of 0.15 m, then

$$F = \frac{Du}{l^2} = 5 \cdot 10^{-3} u \text{ Newton.} \tag{3}$$

The use of a horizontal torsional wire is only possible if the weight of the system is of the order of some mg. C u r i e attached the sample in A. F e r e d a y and W i e r s m a (1935), measuring anisotropies, attached it in C. In the apparatus used by us it is fixed in B.

c. F a r a d a y introduced the normal balance, afterwards used by P a s c a l (1910). A horizontal beam is lying loosely upon a knife (fig. 4). The carrier at the right side often is somewhat long,

0.75 m. The magnetic force works in a vertical direction while the rotation of the beam is observed (see also d e H a a s and v a n A l p h e n, 1930).

d. In the balance of S u c k s m i t h (1929) the sample is suspended on a phosphor bronze ring (fig. 5). A displacement of the sample causes a deformation of the ring, which can be examined by means of two mirrors placed at the optimum positions. Rapid measurement is possible.

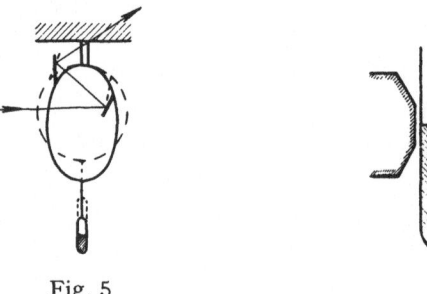

Fig. 5   Fig. 6

e. In Q u i n c k e's m e t h o d (1885) a liquid balances in a U-tube (fig. 6). Here no empty carrier has to be taken into account, but surface tension effects are introduced. The weight of the liquid may be used for compensation, being likewise continuously variable. The sensitivity is enlarged by using a more or less inclined tube or, in order better to observe displacements of the liquid, by applying a constriction in the tube. Gases are often measured with this type (see also W i l l s and B o e k e r, 1923 and fig. 12).

The zeropoint can be fixed by means of a scale-mirror system, in which the mirror makes a rotation, directly coupled with the system (*b*, *c* and *d*) or coupled with a transmission if the system suffers displacements (a_2); also by means of a microscope (a_1 and *e*). In many magnetic measurements the temperature plays an important part. In thermostats most often liquids are used, as their specific heat is large and the convection needed for temperature equilibrium can easily be augmented by stirring, while the thermal contact with the substance is rather good. Liquids can easily be applied in systems with a vertical carrier. In other apparatus most often gas thermostats are built.

Surveying the apparatus mentioned before, we do not apply *a*, as according to (2) and (3) this instrument, on measuring liquids, is less sensitive than *b*. Q u i n c k e's method involves surface tension effects, though they can be avoided. As concerns the balances, the less the gravitational force is used in the apparatus, the more stable and rapid acting it becomes. We have chosen the *torsion balance b*, as it combines *stability* and *sensitivity*.

Magnetic part

Electromagnets are often used. Their fields are limited by the saturation of the iron core or the heating-effect of the current. Permanent magnets also can be applied. Here the apparatus has to be removed to find the zero-point. To obtain a big reproducible force, several methods can be followed, of which we shall describe some.

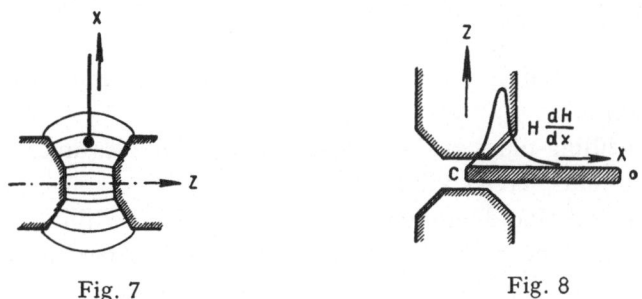

Fig. 7 Fig. 8

a. F a r a d a y (1855) took a sample, small compared with the size of the field, and measured the force at the maximum of $H \, \partial H / \partial x$ (fig. 7). In the case of an ordinary electromagnet the gradient is strongest perpendicular to the field. According to (1) and I (8)

$$F_x = (\mathbf{M} \cdot \text{grad}) \, H_x = v\mu_0 \, (\varkappa_1 - \varkappa_2) \, H_z \, \frac{\partial H_x}{\partial z} =$$

$$= v\mu_0(\varkappa_1 - \varkappa_2) \, H_z \, \frac{\partial H_z}{\partial x} \approx m\mu_0 \chi_1 \, H_z \, \frac{\partial H_z}{\partial x} , \quad (4)$$

in which $\varkappa_1$ is the susceptibility of the sample. In connection with the presence of a surrounding medium of which the susceptibility is $\varkappa_2$, the formula contains the factor $\varkappa_1 - \varkappa_2$. If $\varkappa_1$ depends upon the field, this method has advantages. We only need small quantities of our compounds, but then obtain only relatively small forces. In the neighbourhood of the maximum of $H \, \partial H / \partial x$ the dependence of F on the place is smaller but still considerable.

b. G o u y (1889) used a long rod, formed from the substance, reaching from the centre of the field outwards (fig. 8). Here we find,

2

integrating (4) from the centre of the field outwards, for the total force exerted upon the rod

$$F_x = \int_c^o S\mu_0(\varkappa_1 - \varkappa_2)\, dx\, H_z \frac{\partial H_z}{\partial x} =$$

$$= \tfrac{1}{2} S\, \mu_0\, (\varkappa_1 - \varkappa_2)\, (H_c^2 - H_o^2) \approx \frac{m}{2l}\, \mu_0\varkappa_1\, (H_c^2 - H_o^2), \quad (5)$$

where S, the surface of a cross-section of the rod, is assumed to be constant all over its length (l). Q u i n c k e's method is always one of this type.

This gives the following advantages:

1. H_c, the field strength near the centre of the magnet, depends only slightly upon the place while H_o is small.

2. The force is much bigger than in case a.

3. The effect of the empty carrier is small in comparison with that of the sample.

4. The filling-plugs at the right side and air bubbles, in the case of volatile liquids, have no influence.

On the other hand $\varkappa$ may not depend upon H, the filling must be homogeneous and S constant, especially in the steeper parts of the $H\,\partial H/\partial x$ curve (see III.3).

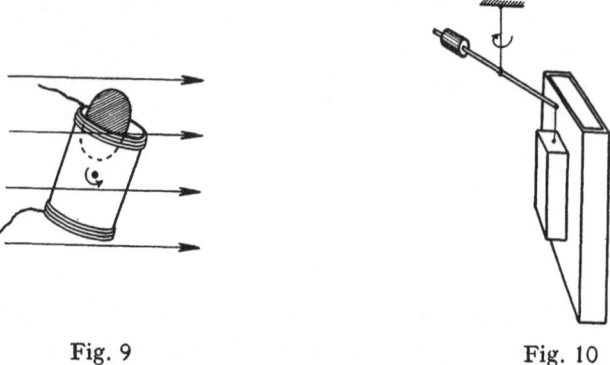

Fig. 9   Fig. 10

c. In both foregoing methods displacements in non-homogeneous fields are used. We can also apply rotations in homogeneous fields. S c h u l t z (1939) used a couple-balance, consisting of a thin-walled tube, carrying two sets of windings at the upper and lower end (fig. 9). The tube can swing around a horizontal axis. If the two sets of windings are bearing opposite currents, the two couples, exerted upon the balance, will compensate each other. The unknown substance can be placed in one of the sets and so give an asymmetry, which can be compensated by changing the current in one of the coils.

This method is a so-called magnetometer method, often used for ferromagnetic substances. The forces, exerted by the fixed sample upon the magnetising agency, are measured. Here we have the advantage that the substance can be replaced without affecting the measuring system. A disturbing effect is the earth field, though its influence can be decreased by applying astatic systems (as e.g. in S c h u l t z' method).

R a n k i n e (1934) also developed a simple and sensitive magnetometer for measurements of small susceptibilities (fig. 10). He suspended a rod-shaped steel magnet on a torsion balance at a small distance of a cuvette with the substance to be examined. The forces upon the rod are measured.

d. K r i s h n a n (1935) measured the frequency of oscillation of magnetically anisotropic material in a homogeneous or isotropic material in a non-homogeneous field. The frequency is determined by the magnetic and torsional moment, working upon the system; the latter is eliminated, measuring with and without field. The moment of inertia is measured, adding a known moment of inertia.

We had a C a r p e n t i e r electromagnet at our disposal. The field is about $1.2 \cdot 10^6$ A/m (15 kilogauss), the distance of the pole-pieces about 10 mm. As we measured liquids, available in sufficient quantities, we choose the G o u y method.

Measurement

It is of importance to make the measurement independent of variations of the currents supplying the magnet and compensating system. This often can be done by applying a compensating apparatus with properties similar to those of the measuring system so that a null-method can be obtained in this respect. Thus A h a r o n i and S i m o n (1929) used a balance with two magnetic substances each in a coil, carrying the same current (fig. 11). One sample was kept under the variable conditions to be studied, the other under constant conditions, while the coil surrounding the latter sample had a variable position with respect to it. Thus the arrangement becomes rather symmetrical and large currents during a short time can be used.

Working with a compensator with two attracting coils, we can make the quantity $H \, \partial H / \partial x / I^2_{comp}$. more independent of the magnet current if the compensating current is shunted off from the main magnet current, particularly if the magnet is far enough from saturation.

When a small effect has to be measured, it will be advantageous to compensate the main effect directly. In the F a r a d a y method, according to (4), we have the possibility of varying $\varkappa_2$, the susceptibility of the surrounding medium, or $\varkappa_1$ of the movable part if the surrounding medium has to be measured, until $\varkappa_1 - \varkappa_2$ has become small. Also if solid bodies of irregular form have to be measured, we can do so and afterwards measure the susceptibility of the solution used as surrounding medium.

Fig. 11   Fig. 12

In the G o u y method we can symmetrise the cuvette and fill the opposite side with a substance of average susceptibility. Following Q u i n c k e's method, W i l l s and B o e k e r (1932) applied two connected tubes (fig. 12).

In our torsion balance we apply the G o u y method with a double cuvette (see III fig. 2). One side is filled with the unknown substance, the other side with the compensating liquid, always kept the same. By doing so the main part of the diamagnetic effect, of which the spread is much smaller than the effect itself, is compensated; moreover a change of the density due to a variation of the temperature has less effect. The differences in the substances are measured with a current-bearing conductor, situated around the sample. This current is taken off from the main magnet current (see III). We reached a precision of 1 pro mille of the diamagnetic effect, while a measurement is performed in 25 minutes.

3. Induction methods

Integrating I (2) we get

$$V = -\frac{d}{dt} \int (\mathbf{B} \cdot d\mathbf{S}) = -\frac{d\Phi}{dt}, \qquad (6)$$

where V is the voltage generated in a magnetic field, and Φ [1]) the flux threading the surface S of the coil, multiplied with the number of turns placed in the field. If this field is caused by currents, it can be calculated by applying I (1) and I (4). For a definite arrangement of coils etc. it follows that

$$\Phi = MI = (1 + f\varkappa) M_0 I, \qquad (7)$$

where M is the coefficient of mutual induction in that case. It can easily be proved that M is the same if we exchange the coil which generates the magnetic field (the primary) and that in which the e.m.f. is induced (the secondary coil). f is called the filling-factor.

Now (6) becomes:

$$V = -M\frac{dI}{dt} - I\frac{dM}{dt}. \qquad (8)$$

M contains the properties of the medium ($f\varkappa$) and hence indicates changes caused by our substance. As (8) shows, there are two possibilities of measurement: *changing the current I* or *the coefficient of induction M*, in other words the geometry of the arrangement. We shall discuss them apart. According to literature on susceptibility measurements, only the first method has hitherto been used.

Varying current

As concerns the frequency of the applied current, we can distinguish three cases: the b a l l i s t i c, the l o w and the h i g h f r e q u e n c y method.

a. If we measure b a l l i s t i c a l l y, we reverse the direct current I once. The induced voltage acting on a ballistic galvanometer causes a deflection depending only upon the total change of

---

[1]) Hereafter we shall distinguish φ, being the number of lines of force threading a surface, and Φ, the quantity occurring in (6), i.e. φ multiplied with the number of turns, or more precisely its effect integrated over the whole secondary coil.

the magnetic flux (see e.g. d e H a a s and W i e r s m a, 1935; fig. 13).

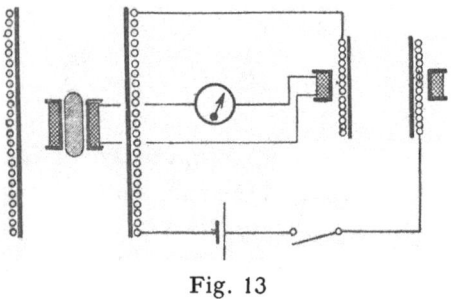

Fig. 13

b. The application of a l t e r n a t i n g c u r r e n t has the advantage that amplifiers with a less sensitive galvanometer or oscillograph can be used. To improve the apparatus one will apply compensation or use a bridge, so that a null-instrument with respect to the current and sometimes also with respect to the frequency is obtained.

Though self-inductance measurements have been performed, it is easier to apply a mutual inductance system (see e.g. C a s i m i r, 1939). Then two coils are used, a primary and a secondary, each of which can fulfill a part of the requirements of the measurements; furthermore one can apply two oppositely connected secondary coils so that many disturbances are eliminated (fig. 14). One of the secondaries can be filled with the sample.

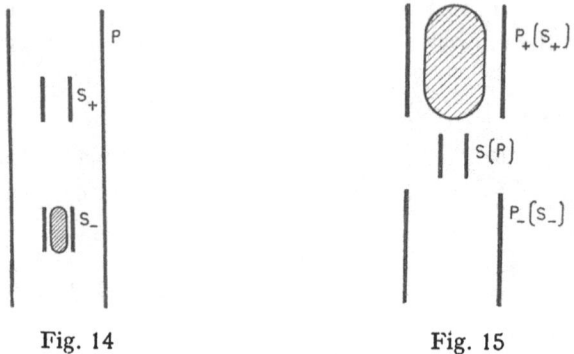

Fig. 14   Fig. 15

A system different from this contains two oppositely connected primary coils P_+ and P_- and one secondary S placed between them (fig. 15). Here one of the primaries is filled with the sample. It is an

advantage that the voltage across the secondary coil is mainly
compensated. The discussion of this arrangement is simplified by
exchanging the primary and secondary coil. Then again a secondary (S_+) is filled with the sample, but it is now placed in the stray
field of a small primary coil (P). This means that small displacements of this coil will have large effects. Furthermore M has
become small as the coils are placed rather far apart. Therefore we
applied the system shown in fig. 14.

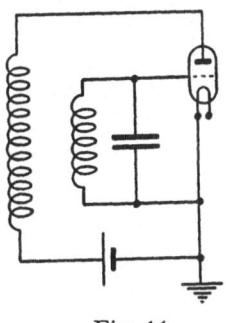

Fig. 16

c. A third system is based on the application of the h i g h   f r e-
q u e n c y   g e n e r a t o r   (fig. 16). One uses an oscillating system,
consisting of a coil filled with our sample and a condensor; thus the
measurement is reduced to a frequency measurement $(1/\nu = 2\pi\sqrt{LC})$.
The comparison with a second oscillating system, kept under constant conditions, furnishes a difference-frequency in the acoustic
region, which can easily be examined (see e.g. G o r t e r   and
B r o n s, 1938).

In the region of high frequencies this system is simpler than a
bridge, but ν is also affected by the dielectric properties of the sample
in connection with the self-capacity of the coil and the damping, due
to the change of the losses in the circuit. As the radio frequency
oscillator cannot be trusted to give us the required reproducibility
of $3 \cdot 10^{-9}$ (see below), we chose a mutual inductance system.

We give a short description of this *mutual inductance apparatus*:
If we use a primary coil with n turns per m, carrying a current

$$I = I_0\, e^{i\omega t}, \tag{9}$$

and a secondary coil with N turns, placed inside the primary and having a surface S, it easily follows from (8) (see also IV.1) that

$$V_{eff} = - i\omega M I_{eff} \tag{10}$$

and

$$M = (1 + f\varkappa) \mu_0 nNS. \tag{11}$$

M can be increased by using coils with a great number of turns. Now the presence of our sample ($\varkappa \approx - 700 \cdot 10^{-8}$) will give a variation of M

$$\Delta M = f\varkappa M_0 = - 3 \cdot 10^{-6} M_0, \tag{12}$$

where f, the filling-factor, in practice is about 0.5.

Requiring a precision of 1 pro mille of our effect, corresponding to a relative error of $3 \cdot 10^{-9}$ in the voltage, 10 V has to be generated as the disturbing voltage is $3 \cdot 10^{-8}$ V, due to noise effects and errors introduced e.g. by capacitance. This precision can be reached, though the apparatus thereby becomes somewhat large. The exchange of the sample is rather easy and in 8 minutes a measurement can be performed. Iron cannot be used in this apparatus as eddy currents and ferromagnetic effects are too large (with an apparatus, described in V.3, this was investigated). According to literature until now the method was only applied for measurements on paramagnetic substances, having much larger susceptibilities.

Varying coefficient of induction

In this method we apply a constant primary field and move our sample with respect to it. In the inductance apparatus with varying current the change in the spread of the lines of force, caused by our sample and threading the secondary coil placed in the neighbourhood, alone is of interest. Here in connection with the application of ferromagnetic material also the change of the total number of lines of force throughout the whole circuit is important. We shall consider these two cases apart.

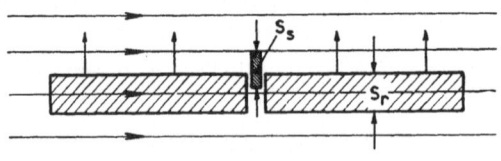

Fig. 17

a. We assume a *homogeneous field* in which a long rod, directed parallel to the lines of force, moves perpendicular to its axis (fig. 17).

At one place there is a slit to allow a coil S, that stays at rest in the magnetic field, to cross the rod. Suppose that both the coil and the rod have a square cross-section so that their common surface, and therefore also the flux threading the coil, increases and afterwards decreases linearly (fig. 18) on moving the rod at a uniform velocity.

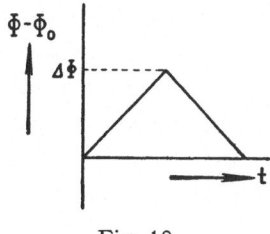

Fig. 18

If the original field is $B_0 = \mu_0 H_0$, then inside the rod we get $B_i = \mu_0 H_0 (1 + \gamma \varkappa)$ so that the jump in B on crossing the rod is $\gamma \varkappa B_0$. For a long rod $\gamma = 1$. In general the maximum change in the flux is

$$\Delta \Phi = f \varkappa \Phi_0, \tag{13}$$

where, like in (12), f again is the filling-factor, about 0.5, and Φ the flux through the coil, multiplied with the number of turns.

b. Next we consider an *iron toroid* (see V fig. 2), containing a piece of permanent magnet steel or a current-carrying coil, furthermore a secondary coil. All lines of force go through the toroid, cross a small air slit and do not shift if something changes (otherwise case a). If we move e.g. a paramagnetic (or a ferromagnetic) substance through the slit, this is shortened and therefore the flux in the circuit increases linearly with the surface, passed by the substance (see fig. 18). Formula (13) here again is applicable (see V.1).

As we see, both effects go parallel and therefore strengthen each other.

For a good measurement filtering is essential and hence we must try to obtain a sinusoidal voltage. This means that more impulses of the form given in fig. 18 must follow each other at the right distance, while furthermore the impulses should have a somewhat more suitable form, which indeed occurs in practice. If in the apparatus more secondary coils are used, we can connect them oppositely one after another and thus in general obtain a more symmetrical curve (fig. 19). Of course the frequency now is halved. We have still a second advantage. If a current produces the magnetic field, its fluctuation according to (8) will generate voltages in the coils:

$$V = -MI\left(\frac{\dot{I}}{I} + \frac{\dot{M}}{M}\right),$$

and hence the amplitude of the ripple of the current in the frequency band used has to be decreased down to $3 \cdot 10^{-9}$. The use of coils oppositely connected gives a possibility for this. In the measuring-generator, as we shall call the sort of apparatus described in this paragraph, one obtains an alternating voltage in connection with a periodicity in space, while the alternating current has a periodicity in time and therefore they can be separated. In further calculations we shall assume the application of this alternation in the position of the coils.

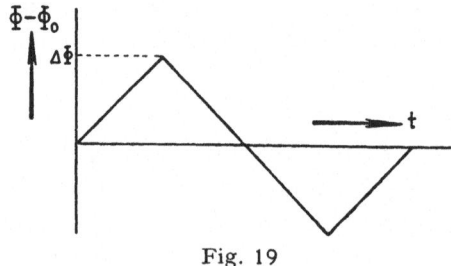

Fig. 19

If we make a F o u r i e r analysis of the curve in fig. 19, the main term is

$$\frac{8}{\pi^2} \Delta\Phi \sin \omega t.$$

As case a and b strengthen each other, we assume an amplitude somewhat (0.2) larger:

$$\Phi = \Phi_0 + \Delta\Phi \sin \omega t. \tag{14}$$

Now the e.m.f. becomes, referring V_{max} to a single impulse,

$$2 V_{max} = -i\omega \Delta\Phi = -f \times \cdot i\omega \, \Phi_0$$

so that

$$V_{eff} = -f \times \cdot i\omega \frac{\Phi_0}{2\sqrt{2}} = -f \times \cdot i\omega \, M \frac{I}{2\sqrt{2}}. \tag{15}$$

To obtain a further knowledge of f we can apply (13) and calculate the maximum difference of the flux; the values of f will not exceed 0.5. It is obvious that V is mainly determined by M.

Now (15) has to be compared with (10) and (12). At a glance we see that the effect generated is directly proportional to x just as in the force method. In the induction method with varying current this cannot easily be achieved, without the disturbances still being proportional to the main effect. As will be seen in IV, dielectric and eddy current effects are the principal trouble in this method. In the measuring-generator these effects do not occur, furthermore the application of iron cores increases the field strength by an order of magnitude. Also permanent magnet steel, without energy dissipation, can be used. On the other hand the motion introduces vibrations, the consequences of which here become the limiting factor, and also no moving metal can be allowed in the neighbourhood of the coils.

In connection with the great advantages we built a trial apparatus. It was fitted up in a very simple manner but gave good results: 8 pro mille of the diamagnetic effect was resolved directly. This precision, however, could not be kept up in more suitable apparatus built later, one with moving sample, the other with moving magnets. Here no more than 0.1 of the diamagnetic effect, $\Delta x = 1 \cdot 10^{-6}$, could be seen, the disturbing voltage being about $1 \cdot 10^{-5}$ V ($3 \cdot 10^{-8}$ V in the apparatus with varying current). Though no diamagnetic measurements were performed, the apparatus will be described for completeness sake in V. A more solid construction will possibly yield better results. Furthermore an absolute ballistic measurement based upon these principles and an analogous dielectric apparatus will be considered.

REFERENCES

A h a r o n i, J. and F. S i m o n (1929). Z. phys. Ch. (B) **4**, 175.

C a s i m i r, H. B. G., W. J. d e H a a s and D. d e K l e r k (1939). Comm. Leiden 256*a*.

C u r i e, P. and C. C h é n e v e a u (1903). J. Phys. (4) **2**, 796.

F a r a d a y, M. (1855). Exp. Res. **3**, 27 and **497**.

F e r e d a y, R. A. and E. C. W i e r s m a (1935). Comm. Leiden 237*a*.

G o r t e r, C. J. and F. B r o n s (1937). Physica **4**, 579.

G o u y, L. G. (1889). C. R. **109**, 935.

H a a s, W. J. d e and P. M. v a n A l p h e n (1930). Comm. Leiden 212*a*.

H a a s, W. J. d e and E. C. W i e r s m a (1935). Comm. Leiden 236*b*.

K r i s h n a n, K. S. and S. B a n e r j e e (1935). Phil. Trans. (A) **234**, 265.

P a s c a l, P. (1910). C.R. **150**, 1054.

Q u i n c k e. G. (1885). Ann. d. Physik **24**, 369.

R a n k i n e, A. O. (1934). Proc. Phys. Soc. **46**, 1 and 391.

S c h u l t z, B. H. (1939). Comm. Leiden 253*d*.

S u c k s m i t h, W. (1929). Phil. Mag. **8**, 158.

W e i s s, P. and G. F o ë x (1911). J. Phys. **5**, 1 and 275.

W i e r s m a, E. C., W. J. d e H a a s and W. H. C a p e l (1930). Comm. Leiden 212*b*.

W i e r s m a, E. C. and H. R. W o l t j e r (1929). Comm. Leiden 201*c*.

W i l l s, A. P. and G. F. B o e k e r (1932). Phys. Rev. **42**, 687.

CHAPTER III

THE TORSION BALANCE

1. Magnet

In fig. 1 the system [1]) is reproduced. The magnet has to be placed vertically as rotations around a vertical axis occur. We use a C a r-p e n t i e r magnet, watercooled at the upper and lower ends of the coils. The maximum continuous load amounts to 7.5 A, limited by the heating of the coils. The fields, always increasing the current up to the right value, can be reproduced rather well on approaching saturation. The diameter of the iron core is 70 mm, the distance of the pole-pieces 10 mm. The whole apparatus has been placed under a box of cardboard to prevent air currents from disturbing the adjustment.

In 3 the arrangement is considered quantitatively. The G o u y method (II.2) is applied. To avoid large mistakes we measure at two field strengths: $1.15 \cdot 10^6$ A/m and $1.25 \cdot 10^6$ A/m (15.6 kilogauss, 7.5 A). The magnet current is examined by means of a compensator, measuring the drop of potential across a normal resistance of 0.1 Ω. In the latter series the inaccuracy of the field scarcely contributed to the error in the measurement.

In the first series (1944) the current supply was furnished by a $2 \cdot 110$ V dynamo of which half the voltage was used in a regulating system, keeping the current constant at some 0.1 pro mille so that hysteresis effects, caused by the ripple of the voltage, could not affect the value of our field. In the second series (1945) a battery of 120 V has been used. Its constancy was sufficient when the battery was only used by us; furthermore the direct magnetic compensation of the main effect, described at the end of II.2, was applied so that the disturbance of the field is much less.

---

[1]) This apparatus was suggested by dr. P. v a n d e r L e e d e n and first built by ir. H. G. K l i n k e r t. The author simplified the apparatus and performed the measurements described in VI. Also the valuable remarks of dr. P. v a n d e r L e e d e n on other parts of this thesis should be mentioned here.

Temperature regulation

The pole-pieces are covered by a flat copper box in which stream-
ing water,heated by a hand-regulated current, furnishes the temperature of 20°C. The space between the coils has been screened off with
cardboard (C) to obtain a chamber of constant temperature. If deviations occur, the density changes about 1 pro mille per °C. As the
compensating liquid also has a thermal expansion and the time-
constants of both halves are equal, the correction is negligible, even
if the temperature difference is of the order of 0.5°C. For the calibration points with air and water the correction still has to be applied.

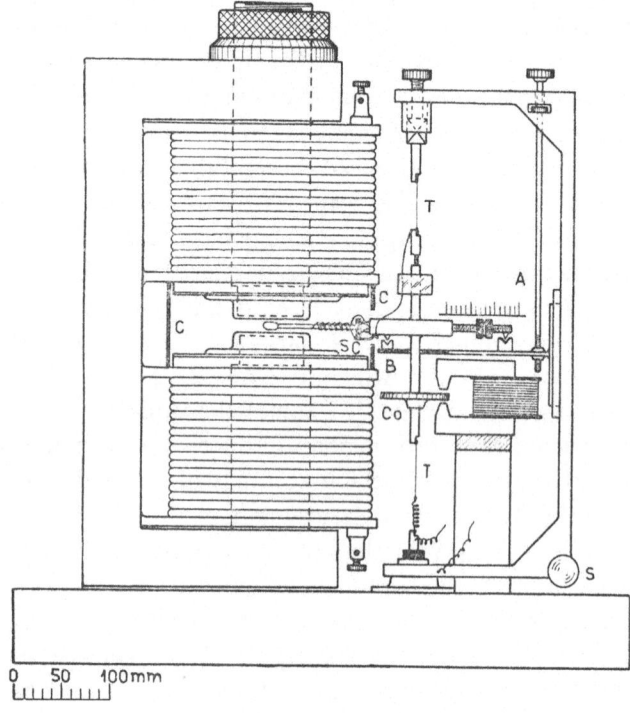

Fig. 1

In the first series we waited 20 minutes after filling the cuvette.
During this time, the temperature could adjust itself; moreover the
sample was magnetised to clean the liquid from iron impurities as
much as possible. In the second series, after pouring the liquid through
a funnel in the field of a small magnet (see VI) and measuring in a

room with a temperature of about 20°C, it is sufficient to wait 5 minutes. The whole measurement then takes 25 minutes, the filling included.

2. Balance and measuring system

The movable parts of the balance are constructed of copper to limit ferromagnetic effects. Glass is used for those parts which move in the magnetic field to prevent too strong a damping.

We apply a magnetic damping system, consisting of a copper disk (Co), with a thickness of 3 mm and a radius of 60 mm, which moves in the field of a small electromagnet ($\approx 1C^5$ A/m ≈ 1 kilogauss). The critical damping is established by means of the field current. The glass cuvette (fig. 2) is taken out as a whole when we have to fill it. The length of the cuvette is $2 \cdot 75$ mm, the cross-section $11 \cdot 6$ mm^2, the thickness of the wall 0.8 mm. Two sliding-contacts (SC) furnish the current supply for the single turn, laid around the compensating cuvette. By means of an arresting system (A), the torsion fibre (T) can be relieved, the movable system then resting upon the horizontal beam (B).

The torsion fibre (T) has been taken so thick that 1 pro mille of our effect gives a deflection of 1 mm at 1 m distance. The thickness of the fibre ($1.1 \cdot 0.06$ mm^2) could have been decreased much more without the fibre breaking at the weight of about 200 g which it had to carry. However, the increase of sensitivity thus attainable is of no use as the adjustment would cost much more time. Now the time to reach the zero-point is about 15 sec for a deflection of 0.1 m (0.1 of our effect). Furthermore the constancy of the zero-point would decrease, both by the thinness of the wire and the increase of the time of measurement. In practice it appears that the adjustment would then have to be repeated more often.

The reproducibility of the zero-point during a series is of course of great importance. In the former experiments some turns, carrying a current and placed in a central position upon the cuvette, furnished us a fixed zero-point in the centre of the field. Afterwards nothing except the compensating coil was placed upon the cuvette and we then trusted upon the fixed position of the balance and scale. The zero-point can be established by turning the frame with the screw S.

A voltage difference is taken off from the magnet current by means of a normal resistance. Through an amperemeter, a decade resistance

and the upper and lower torsional wire a current ($I = 4$ or 20 mA) is
supplied to the single turn, fixed to the compensating part of the
cuvette. This has the advantage that if the magnet current changes
somewhat, according to (4), the variation of I will be compensated
by that of H. In this manner the original effect is halved. In the first
series a resistance was read off and after correcting with extra
resistances, $1/R$ was taken as a measure for the effect. In the second
series the current has been read off directly. Only the linearity of the
amperemeter and the current ratio, if the amperemeter is shunted, is
of interest. The latter is necessary in the measurement of the air
point. All our measurements are relative to water.

The symmetry of the carrier gave the possibility to compensate
the main diamagnetic effect. This has been done in the second series,
using a liquid with average $\varkappa$. By doing so the effect to be measured
and hence the rôle of the compensator is reduced, the compensation
then ranging from $+0.2$ up to -0.1 of the total effect. Another
advantage of our method is the small effect due to the carrier
and gas bubbles appearing on the outer side of the cuvette when
the liquid evaporates.

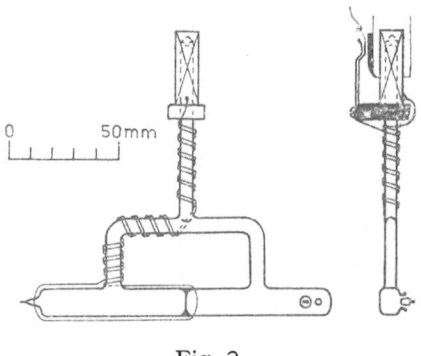

Fig. 2

As follows from a comparison of the measurements performed for
both field strengths, in the second series the effect of the adjustment
and change in temperature is less than 0.2 pro mille. Due to hysteresis and saturation of the magnet we get an error of 1 pro mille in the
adjustment of the air measurement if the current has been made 30
pro mille too high before the right value is adjusted and if an error in
the current of 3 pro mille has been made. In the case of the water
measurement these numbers are 100 pro mille and 15 pro mille

respectively. Still the effect of ferromagnetic substances situated anywhere upon the cuvette can limit the precision. Adding the error caused by the amperemeter and the calibration, the physical error will be about 1 pro mille.

3. Calculation of the forces working upon the system

The *magnetic force* exerted upon the substance according to II (5), adding a term representing the effect of the inhomogeneity of the cuvette (see II (4)), is

$$F = (\varkappa_1 - \varkappa_2)\,\mu_0 \left[\tfrac{1}{2}\,S(H_c^2 - H_o^2) + \int \Delta S\,H_z \frac{dH_z}{dx}\,dx \right], \quad (1)$$

where ΔS is the deviation of the surface from the average value
The principal term of F in our experiment is

$$\tfrac{1}{2}\varkappa\,\mu_0\,SH_c^2 = 3 \cdot 10^{-4}\ \text{Newton}$$

and the moment of this force, using an arm of 0.145 m, is

$$M = 4.4 \cdot 10^{-5}\ \text{Nm.}$$

A deflection of 1000 mm, due to this effect for a scale-mirror distance of 1100 mm, corresponds to a rotation of 0.45 rad. Hence the torsional constant $D = 1 \cdot 10^{-4}$ Nm/rad. For the upper fibre, giving the biggest contribution to D, $l = 50$ mm, the cross-section is $1.15 \cdot 0.06$ mm^2 and the torsion modulus $G = 0.45 \cdot 10^{11}$ N/m^2, so that

$$D = \frac{G}{l}\,\tfrac{1}{3}\,ab^3 = 0.75 \cdot 10^{-4}\ \text{Nm/rad.}$$

Now the moment of inertia is: $J \approx 1 \cdot 10^{-3}$ kgm^2. The total weight is about 0.2 kg. There follows then that the period $T = 2\pi\sqrt{J/D} \approx$ 18 sec, which fits with the experiment.

The *compensating force* is the L o r e n t z force; it is given by

$$F = \mu_0\,Ib(H_c - H_o) = 3 \cdot 10^{-4}\ \text{N} \quad (2)$$

and exerted upon the single turn, carrying a current of 18 mA; the length of its short sides $b = 11$ mm (fig. 2). The longer sides of the turn have no effect, if they are strictly rectilinear; the supply wires have been twisted. The resultant moment is

$$M = A(\dot{H}_c^2 - H_o^2) - \mu_0\,r\,Ib(H_c - H_o) - D\frac{x}{r}, \quad (3)$$

3

in which x is the position of the centre of the cuvette and A the abbreviation of some factors (see (1)). In the zero-point ($x = 0$) the equilibrium is established when

$$A(H_c^0 + H_o^0) = \mu_0 \, rIb. \tag{4}$$

Hence

$$x = \text{const} \cdot I \tag{5}$$

provided the same geometry and the same field H_c is retained. The calibration constant is found from an air and a water measurement. Now (3) becomes:

$$M = A(H_c - H_o)(H_c - H_c^0 + H_o - H_o^0) - D\frac{x}{r}.$$

Experiments, concerning the dependence of the field strength upon x, show that according to this formula the position becomes unstable if $x < -80$ mm. This point of instability, however, lies far outside our scale, being $\pm$ 15 mm at the place of the cuvette. Still it reminds us of the possibility of such effects. But these are due to the second term in the formula for the force exerted upon the substance (see (1)), neglected in the derivation above; viz. irregularities of the cross-section S of the cuvette.

The instability actually occurred when we had fused a platinum wire to our cuvette. The cuvette, blown from a factory-made flat tube, was distorted by this process. Therefore we again fixed our turn with shellac.

Of great interest is the *dependence of the adjustment on the choice of the zero-point*. In the measurement with the distorted cuvette this effect amounted to 0.5 pro mille per mm displacement of the zero-point. As the effects for a magnetically compensated cuvette were sometimes bigger than for an air measurement, it followed that the inhomogeneity of the cuvette and not the single turn gave the trouble. In the final apparatus an extremum could be found, not far from the central position; here 10 mm displacement changed the effect by 0.5 pro mille.

During these measurements at different places of the scale we noticed that susceptibility ratios, calculated at a fixed place, were not reproducible over the scale. Differences up to 10 pro mille occurred. It was found that this was due to the difference in the density of the liquids, causing displacements of the carrier. Their maximum value is found with the equation (fig. 3) $u \cdot 0.2 = 0.145 \cdot 3 \cdot 10^{-3}$ so that $u = 2$ mm. Balancing with the compensating weight W, which can easily be done with the naked eye, this effect is eliminated.

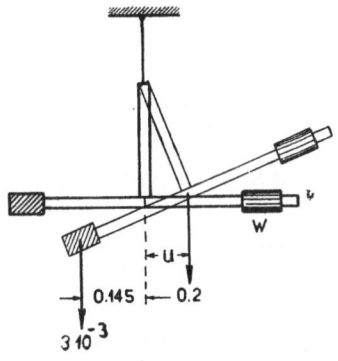

Fig. 3

With the magnet used above strong fields can be applied, but as its size compelled to take a large measuring system, the difficulties resulting made the apparatus less convenient. Therefore one might try to construct a much *smaller instrument*, by using e.g. a permanent magnet. Comparing with the apparatus described above, considerations of similarity show that this will be possible. The moment of the magnetic force decreases about a factor hundred, but the precision of adjustment can be partially recovered on decreasing the torsional constant D. Furthermore in the apparatus above mentioned the precision of adjustment was more than sufficient. As the dimensions become much smaller, the period even decreases somewhat.

CHAPTER IV

INDUCTANCE APPARATUS WITH VARYING CURRENT

1. Survey of the arrangement

In fig. 1 the apparatus used during the measurements is reproduced. Two oppositely connected secondary coils S, with N turns
each and a surface S_s, have been placed in a solenoïd P, the primary
coil, carrying n turns per m (further described in 5D). According
to II (10) the following holds:

$$V = -i\omega MI, \tag{1}$$

and as we shall see further,

$$M = (1 + f x) \mu_0 NnS_s \tag{2}$$

so that

$$\Delta M = f x M_0 = -3 \cdot 10^{-6} M_0. \tag{3}$$

The measurement of 1 pro mille of the effect requires a measurement
of the voltage V with a precision of $3 \cdot 10^{-9}$.

In 2 we shall describe the compensator with which the effect is
measured. The disturbances affecting this measurement are reported
in 3. As a result we can mention that $3 \cdot 10^{-8}$ V is the lowest measur-
able voltage. This means that the drop of voltage across one coil
must be 10 V so that the value of MI is fixed. ($\omega = 1000$ has been
used). The difference caused by our sample is found by measuring
the voltage across the two coils with the substance successively in
each of the coils. The effect thus has been doubled, but as the difference of two settings of the compensator has to be taken, the error is
also doubled. Therefore in the following our quantities will always
refer to one of the secondary coils. It is a matter of fact that swift
measurement is advantageous as nearly all errors increase with time.
The arrangement allows an adjustment in about 5 sec. In fig. 2 the

measuring scheme is reproduced; apart from auxiliary apparatus
(described in 6), it consists of a primary circuit with the solenoïd P

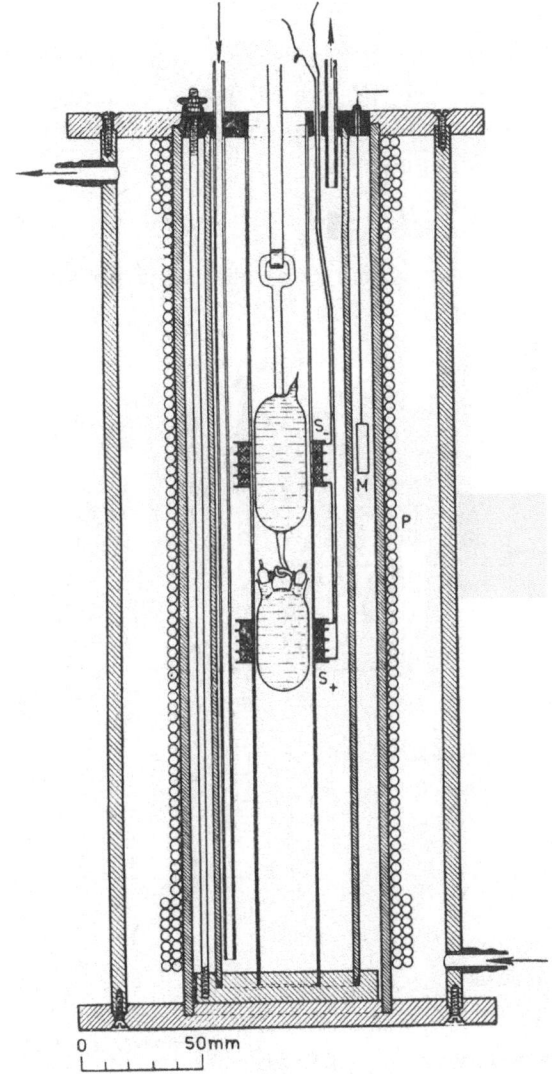

0   50mm

Fig. 1

and two small resistances (*a* and *b*) and a secondary circuit with the
secondary coils and another two resistances (*c* and *R*) of the compensator.

The *calculation of the effect* is mainly based upon the knowledge of j, obtainable with the flux

$$\Phi = \int \varphi \, dn = \frac{N}{l} \int B_x \, dS_s \, dx. \tag{4}$$

When we have an ellipsoïd with one of its axes parallel to the homogeneous main field (H_0), calculation shows that the inner field is a homogeneous one. For an ellipsoïd of revolution the field strength inside the ellipsoïd is given by

$$H^i = H_0 - \alpha \frac{J}{\mu_0} = H_0 - \alpha \varkappa H^i.$$

We shall calculate two special cases, viz. an infinitely long cylinder $(\alpha = 0)$ and a sphere $(\alpha = \frac{1}{3})$.

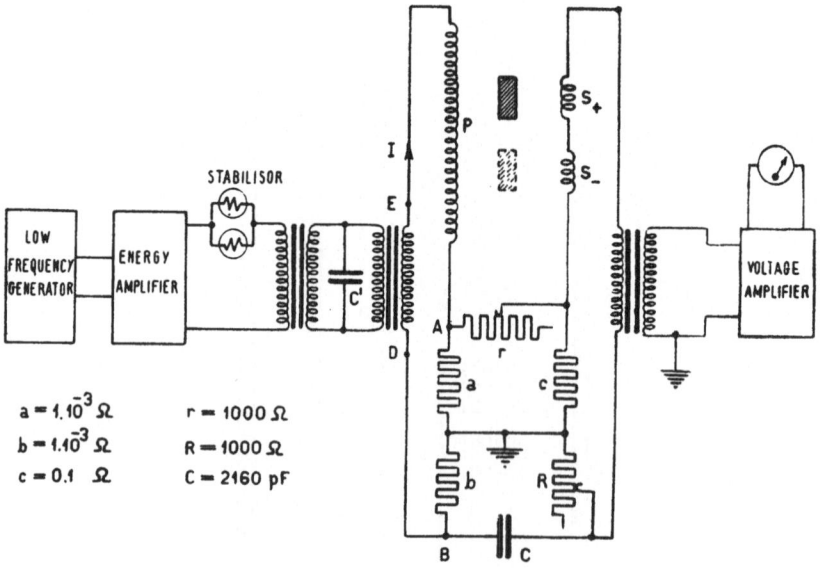

$a = 1.10^{-3} \, \Omega$ $r = 1000 \, \Omega$
$b = 1.10^{-3} \, \Omega$ $R = 1000 \, \Omega$
$c = 0.1 \, \Omega$ $C = 2160 \, pF$

Fig. 2

For a *long cylinder* (fig. 3)

$$H^i = H_0$$

and hence

$$B^i = B_0 (1 + \varkappa).$$

If S_s is the surface of the secondary coil and S_r of the cylinder, then

$$\varphi = (S_s - S_r) B_0 + S_r B_0 (1 + \varkappa),$$
$$\Phi = NS_s B_0 \left(1 + \frac{S_r}{S_s} \varkappa\right). \tag{5}$$

For a *sphere* (fig. 4 gives a cross-section through the axis parallel to the main field)

$$H^i = H_0 - \tfrac{1}{3}\varkappa\, H^i$$

so that

$$B^i = B_0\,(1 + 2\beta),$$

where

$$\beta = \frac{\varkappa}{3 + \varkappa} \approx \tfrac{1}{3}\varkappa.$$

Now we integrate over a cross-section C, perpendicular to the axis of the coil and cutting the sphere, of which the centre O lies upon the axis.

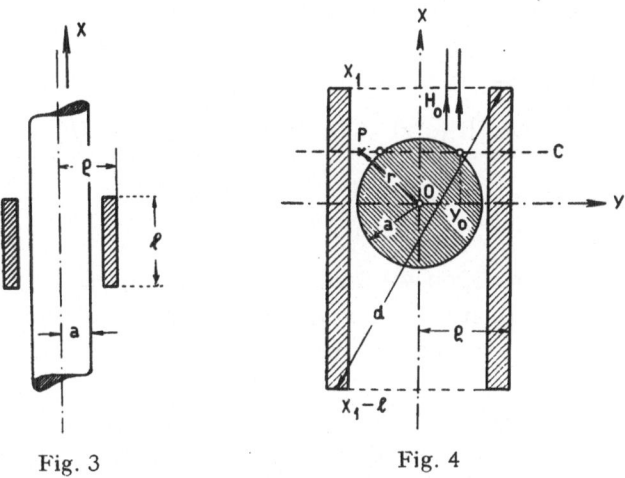

Fig. 3   Fig. 4

Obviously

$$\varphi^i = \pi y_0^2 B_0\,(1 + 2\beta).$$

Outside the sphere we have

$$H^o_x = \mathrm{grad}_x\left\{1 - \beta\left(\frac{a}{r}\right)^3\right\}(\mathbf{H_0 r}) = \left\{1 - \beta\, a^3 \frac{r^2 - 3x^2}{r^5}\right\}H_0,$$

in which the second term represents the effect of the sphere acting as a magnetic dipole. The calculation of φ^o between the sphere and the coil is easy.

Adding φ^i and φ^o, two terms containing $\varkappa$ compensate each other, leaving only the outer limit given by the secondary coil:

$$\varphi = B_0\left\{\pi\varrho^2 + 2\pi\,\beta\,a^3\,\frac{\varrho^2}{(\varsigma^2 + x^2)^{3/2}}\right\} = S_s B_0 + v_{sph}\,\varkappa\, B_0\left\{\frac{\varrho^2}{2(\varrho^2 + x^2)^{3/2}}\right\}.$$

The factor between the braces represents the magnetic field in the centre of the sphere, produced by one turn of the secondary coil. This turn should lie in the cross-section over which the integral is extended. Hence

$$\Phi = NS_sB_0\left(1 + \frac{v_{sph}K_s}{S_sN}\varkappa\right),\tag{6}$$

where K_s is the field of the whole secondary coil, carrying 1 A, in the centre of the sphere; if the latter lies in the centre of the coil

$$K_s = \frac{N}{d},$$

where d is the diagonal of the secondary coil.

As a result we get

$$\Phi = \Phi_0\,(1 + f\varkappa) = NS_s\mu_0\,nI\,(1 + f\varkappa) = MI$$

and

$$f_{cyl} = \frac{S_r}{S_s}, \quad f_{sph} = \frac{v_{sph}}{dS_s}.\tag{7}$$

In our last construction we used $\rho = 14, a = 11, l = 17, d = 32.5$ mm (see fig. 19); then $f_{cyl} = 0.6^2, f_{sph} = 0.2^7$. As the quantity of our sample is limited, we choose a medium case, in which f will be 0.4^5.

A rough estimate of a more general case can be made by integrating (4) over the volume of the sample (v') inside the coil only:

$$\Delta\Phi = \frac{N}{l}\cdot(1-\alpha)\,\varkappa\,B_0\cdot v' = \frac{(1-\alpha)\,v'}{l\,S_s}\varkappa\,\Phi_0.$$

Now we find for f_{cyl} exactly the same and for $f_{sph} = 0.3^3$.

As is obvious, the formulae contain $\varkappa$ and hence the temperature has to be adjusted. Furthermore (7) shows that it will be advantageous to enclose the sample by the secondary coil as much as possible, as the denominator contains the radius of the coil with an exponent between 2 and 3. If this is always done, f will be rather fixed and V is mainly determined by MI.

2. Compensator

The voltage of the coil is measured following a compensation method. We start with two oppositely connected coils with an equal number of turns. As the surfaces may differ ten pro mille, we add some turns with a precision of one turn. Thus a factor 10^{-3} is gained in the compensation. A further adjustment is reached by deforming a supply wire of the primary coil in the neighbourhood of the coils until the original voltage is reduced by at least a factor 10^{-5}. As this possibly only occurs with the inductive voltage and the eddy currents (see 3f) may still give a considerable voltage in phase with the primary current, a piece of metal (M in fig. 1) is placed inside the primary to compensate the losses. The remainder is examined by means of the compensator.

This compensator requires two phase-constant voltages, 90° out of phase, one for the measurement of the mutual inductance and one to compensate the dielectric and magnetic losses. In fig. 2 we see that the primary currents flows through two resistances a and b. Across these resistances we obtain two voltages in phase with the primary current, which can be branched off by means of potentiometers. The upper part consists of two ohmic elements (r and c), the lower part of a resistance (R) and a condensor (C) with a high impedance, thus furnishing the voltages in phase with and 90° out of phase with the primary current respectively. The adjustment can be reached by altering r and R.

In the lower circuit the voltage across b is Ib and hence across R

$$V_{comp} = Ib \ \frac{R}{b + R + \dfrac{1}{i\omega C}} = i\omega \cdot bRC \cdot I. \tag{8}$$

As $1/\omega C = 0.5 \, \text{M}\Omega$, we can neglect R in (8). Deviations from this supposition will in first approximation give voltages in phase with the primary current. The second order term again is in phase with V_{comp} and gives the factor

$$1 + \omega \, RC \left(\frac{\omega L}{R} - \omega \, RC + 2\alpha \right).$$

The effect of the self-inductance L of R is $\omega L/R < 3 \cdot 10^{-3}$. The dielectric loss α in the air condensor is smaller than $3 \cdot 10^{-3}$ and $\omega \, RC = 10^3 \cdot 10^3 \cdot 2 \cdot 10^{-9} = 2 \cdot 10^{-3}$. Hence all non-linear terms can be neglected as they have an order of magnitude of 10^{-5} in the utmost case.

As a result we see that the combination of the two resistances b and R of $10^{-3} \, \Omega$ and $1000 \, \Omega$ and an aircondensor C of about 2000 pF can furnish us an $M' = 2 \cdot 10^{-9}$ H with a linear scale and small phase deviation. Comparing (8) with (1), we see that both voltages are proportional to the current and the frequency and hence a null-method has been obtained in this respect. Compensating we get

$$\Delta M = f \times M = - bC \,\Delta R. \qquad (9)$$

Keeping the same position, x is proportional to R. The proportionality constant can be found by calibrating with a known substance. In our last apparatus $M = 2.5 \cdot 10^{-4}$ H so that M' is about 3 times our effect; 1 pro mille of our effect can be adjusted by changing R 0.4 Ω.

For control we performed one absolute measurement with a spherical sample; the results agree within twenty pro mille with the results given in the literature. With the knowledge of the direction of winding of the primary and secondary coils the negative sign of the diamagnetic susceptibility has been checked.

As the range of the diamagnetic effect is smaller than the effect itself, we compensate the main effect with a second sample. We use a second cuvette with bromine, which has a large diamagnetic effect and can easily be cleaned from iron. This cuvette gives about twice the effect of the unknown sample Sa (fig. 5). In the first measurement (*a*) Sa is in the upper coil, in the second part (*b*) it is replaced by bromine and removed to the lower coil. By doing so the difference of the voltages is reduced.

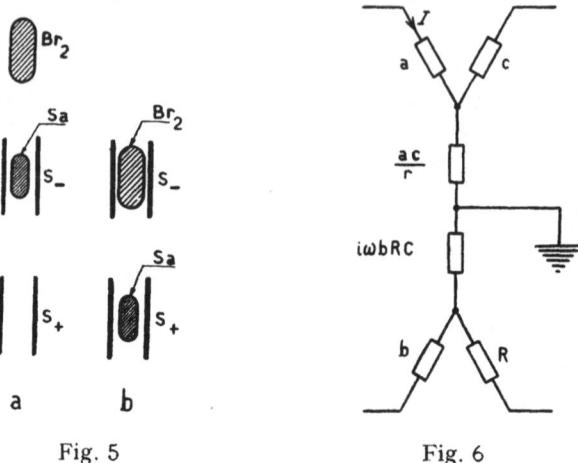

Fig. 5   Fig. 6

The resistance R, the value of which furnishes us the susceptibility, consists of two resistances in series, one to adjust the zero-point and a second to measure the effect. The second resistance is periodically shortcircuited by means of a relais, depending on whether the sample is in the lower or upper coil. In the same way r consists of two parallel resistances.

The upper part of the compensator gives a voltage independent of the frequency, while the losses contain ω and ω^2; still the arrangement is better than others, for which M also has this order of magnitude. As can easily be seen by star-triangle transformation of the compensator (fig. 2 and 6), commutation at A and B is possible. It is only applied for the compensation of the losses. In our scheme fairly low resistances can be used so that the coils are directly earthed by earthing the centre of the measuring scheme. This necessitates the placing of an input transformer in our amplifier, but then also the noise effects of the amplifier become of less influence.

3. Disturbing influences affecting the measurement of the voltage

The coils are the central part of the apparatus and the main difficulties arise from them. In this paragraph we shall investigate how far the simple equations (1) and (2) really describe the situation and what secondary effects are present.

In the secondary coil *noise effects* occur (*a*). Apart from this, *stray fields* from other parts of the apparatus may have an influence (*b*). As the voltage generated in the secondary coil is proportional to the *surface* of the coil, a very constant temperature is required (*c*). The electric field caused by this voltage gives effects in connection with the *self capacity* (*d*).

The assumption of the *homogeneity* of the magnetic field is discussed (*e*). The linearity of the field strength with the current, which is the basis for the measurement, is affected by *eddy currents*, generated in metallic parts present in the magnetic field. Especially the dependence of this effect on the temperature is discussed (*f*). Small fragments of *ferromagnetic material* give non-linearities too (*g*). *Electrolytic effects*, caused by water in the cooling circuit, can also give trouble for this reason (*h*). Finally the primary coil has an electric field in its surroundings which causes voltages in the secondary coil, described by the *mutual capacity* (*j*). *Dielectric losses* may be large (*k*).

a. The thermal agitation of the conduction electrons gives rise to a *fluctuating voltage* (see e.g. M o u l l i n, 1938). We can easily get an insight into this question if we consider a circuit, consisting of a pure resistance R in series with a condensor C (fig. 7). The fluctuations e in R cause voltages V across C:

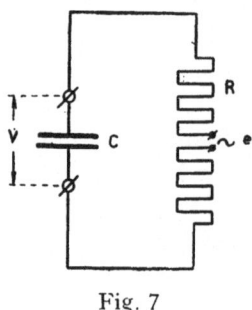

$$V = \frac{e}{1 + i\omega\,RC}.$$

The electrical energy connected with this is

$$C\overline{V^2} = C\int_0^\infty \frac{e^2}{1 + \omega^2\,R^2C^2}\,d\nu = \frac{e^2}{4R}.$$

Fig. 7

As the fluctuations are spontaneous, we add the intensities. Integrating we assume e to be independent of ν, since otherwise the integral would depend upon C. This, however, certainly is not the case as $\frac{1}{2}C\overline{V^2} = \frac{1}{2}kT$ corresponds with one degree of freedom. Furthermore the integral just converges. Hence

$$e\,\Delta\nu = 4R\,kT\,\Delta\nu = 1.5\cdot 10^{-20}R\,\Delta\nu. \tag{10}$$

In our apparatus $\Delta\nu \approx 1$ and $R \approx 50\ \mathrm{k\Omega}$ so that

$$\sqrt{e^2\,\Delta\nu} = 3\cdot 10^{-8}\,V_{eff}.$$

In all formulae effective values are used.

In vacuum tubes we also get statistical fluctuations; in penthodes e.g. the electrons can choose to go to the anode or the screengrid, giving another statistical fluctuation. This effect can be decreased. The variation of the instant of emission of the electrons gives rise to the shot effect. By the use of an input transformer these effects are reduced.

The value of $3\cdot 10^{-8}$ V was imposed as a limit for the other disturbances.

b. To investigate the influence of the *current supply*, we shunt between D and E (fig. 2) a resistance across our coil and measuring system. This can easily be done with a transformer (1 : 25). Placing 100 Ω across the transformer, it is equivalent with 0.2 Ω. Keeping the same current through the coil and doubling the currents in the supply part, a difference in the adjustment of 10^{-2} of our diamagnetic effect is found. Placing the transformer as a series resistance at E, an effect smaller than $5\cdot 10^{-2}$ of our effect is found. However, doubling the current through the coil also, the difference amounts up to 2 times our effect, which suggests that big non-linearities exist in our coils. Hence the supply part has no important effect on the precision. In the beginning, when we used a frequency generator with long supply wires through the laboratory, the supply part had a big effect. The voltage amplifier has not been affected by the other parts of the apparatus. The effects of the coils have been reduced.

c. According to equation (2) V is proportional to the *surface* of the secondary coil. A relative constancy of S of $3 \cdot 10^{-9}$ corresponds to a relative linear expansion of a solid piece of copper of $1.5 \cdot 10^{-9}$, which can be caused by a rise of the temperature of 10^{-4} °C. It appears that enough constancy can be obtained in the voltage difference of the two secondaries by swiftly pumping oil along the coils and by performing a measurement in some seconds. D e w a r flasks outside and inside the secondary coils are also satisfactory, but eddy currents will occur in the silver coating and the filling-factor is easily reduced by a factor 4. There-fore they have not been applied in the final appa-ratus.

d. For the study of the effect of the *self-capacity* we shall assume a condensor C_s to be placed across the secondary coil (fig. 8). If the generated voltage is V_0, then the voltage across the coil is

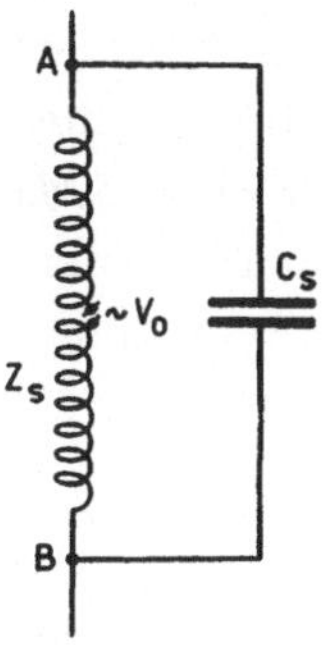

Fig. 8

$$V_{AB} = \frac{V_0}{1 + i\omega\, C_s Z_s} = i\omega\, MI\, (1 - i\omega\, C_s Z_s)$$

and hence

$$V_{self\,cap} = - V_0 \cdot i\omega\, C_s Z_s = V_0 \left(\frac{\nu}{\nu_0}\right)^2 \left(1 + \frac{R}{i\omega L}\right), \qquad (11)$$

where ν_0 is the natural frequency of the coil, Z_s the impedance of the secondary coil. In table I we give some data concerning the several coils that were built before the disturbances mentioned in these paragraphs had been recognised. Four coils are described in 5, viz. A, B, C and D. We give below their natural frequency, the factor $(\nu/\nu_0)^2$ for 175 Hz and in the third line this factor stated in terms of our precision of measurement, being according to (3)

$$\Delta V = fx \cdot 10^{-3}\, V_0 = f \cdot 0.7 \cdot 10^{-8}\, V_0.$$

The second term of the last factor of (11) is given in the fourth line of table I. In table V the absolute value of this factor has also been

taken into account and values of $V_{self\ cap}$ are given for the constructions A, B, C and D.

TABLE I

	A	B	C	D
v_0 kHz	6.5	2.7	130	160
$(v/v_0)^2$	$0.7 \cdot 10^{-3}$	$0.4 \cdot 10^{-2}$	$2 \cdot 10^{-6}$	$1 \cdot 10^{-6}$
$\dfrac{10^3}{fx} (v/v_0)^2$	$0.6 \cdot 10^6$	$0.6 \cdot 10^7$	$0.7 \cdot 10^3$	$3 \cdot 10^2$
$(R/\omega L)_{sec}$	1.6	0.7	0.8	4

As the successive columns of table I and V show, the apparatus has
been much improved. Coil B already stopped being a coil and became a condensor at 2700 Hz. Apart from a slight reduction of the
dimensions of the secondary coil, the main alteration has been the
decrease of the number of turns, from 10,000 up to 1000 roughly.
The self-inductance strongly decreases with it (see table V) as does
also the self-capacity. Moreover the coil has been divided in sections
and papersheets have been laid between the layers (see 4). To keep
the same ωMI, the current, and hence the amount of power needed,
has to be increased.

According to (11) $V_{self\ cap}$ causes a large dependence on the
frequency:

$$\left(\frac{dV}{V}\right)_{self\ cap} = 2\frac{dv}{v}.$$

For the last coil $V_{self\ cap}$ equals our diamagnetic effect so that at first
a constancy of the frequency of the order of 1 pro mille seems to
be required. Besides we have to mention that both coils compensate
each other as concerns this disturbance. A factor 10 or more in the
decrease may be obtained if the coils have been equally wound. Coil
A, consisting of one main and two half coils, became best when we
tuned both halves with an extra condensor until the natural frequency equalled that of the main coil.

After having made the adjustment for the fundamental frequency,
a great number of higher harmonics remain, especially for coil A and
B. They can be filtered out. In connection with the self-capacitance
we use the relatively low frequency of 175 Hz.

e. The *homogeneity* of the primary field is of interest as it occurs in
the coefficient of mutual induction. In the first place the finite
length of the primary coil can cause inhomogeneities up to $2 \cdot 10^{-4}$
of the field strength per mm. The demagnetising field of both poles

can be compensated with small extra coils, wound upon the ends. By doing so a change of less than 10^{-5} per mm can be obtained, which means that a displacement of the secondary coil of 0.3 μ corresponds to an effect 3.10^{-9}. During the measurement it appears that this accuracy can be attained.

Secondly there is the local inhomogeneity connected with the thickness of the wire. The decrease of the effects of capacitance leads to an increase of the current and as the energy provided is limited, the wire of the primary coil has to be taken very thick. In the last coil the diameter was made 4 mm. Several estimates, using the model of circular bands, show that upon the axis the inhomogeneity will then not be large. If we take the mean influence over the length of the secondary coil, which is 4 times the diameter of the wire, the effect will be less than 1 pro mille. As is obvious, in reality our coil is a helix so that upon the axis no inhomogeneities occur.

Corresponding with these effects we must mention the effect of the relative position of the cell, containing the sample and the secondary coil, as this occurs in the filling-factor. In a central position, which is easily found experimentally, the effect is no more than 1 pro mille per mm.

f. The effect of the thickness of the wire, already mentioned in *e*, occurs still more strongly in the *eddy current losses* and the thickness is limited by this effect. The losses arise from the fact that the magnetic field causes voltages in the conductors in the neighbourhood. The currents following from this again induce voltages in the secondary coils. The primary coil forms the main quantity of metal and therefore stranding of several insulated wires to one wire will improve it.

We again have the advantage that the secondary coils compensate each other as concerns the main effect, which compensation is improved by a factor ten on placing a small piece of zinc between the primary and secondary coil. The main effect may be of the order of 1 pro mille of the total voltage, which is of interest as the several parts, compensating each other, may act differently. Such differences may be due to a local temperature rise since the eddy currents, apart from self-inductance, are determined by the resistance, changing 4 pro mille per °C. Changing e.g. the temperature of the upper part of the primary coil with 10^{-3} °C, we get a change of $4 \cdot 10^{-9}$ of the total voltage, which is our precision of measurement.

As the primary coil dissipates nearly all the energy supplied by the energy amplifier (150 W), the cooling is of great importance. The heating is proportional to I^2 and therefore a non-linearity in the dependence of the secondary voltage on the current arises from this. The coolirg of the primary coil of D was improved by applying a swift oilstream.

g. The effect of *iron* of course is considerable. As concerns the main effect it appeared in a special case that a piece of 0.05 mg iron gave 12 times our effect. This means that the adjustment changes 1 pro mille if the 8 g of our substance contain $4 \cdot 10^{-9}$ g, that is a relative quantity of iron of $0.5 \cdot 10^{-9}$. Of course small iron particles may act differently. The above number corresponds to an iron cube with a linear size of the order of 10 μ and with a × of the order of 20.

A second effect is the departure from a straight line of the *B H*-curve, which causes a non-linear effect in the measuring apparatus. It differs from the effects in *f* and *h* as it is instantaneous and thus can be distinguished from them. The non-linearity can amount up to 0.1 of the total ferromagnetic effect. A sudden increase of higher harmonics indicates the presence of iron. The effect is measured, following the method described in *b*, and appears to be of the order of our diamagnetic effect (doubling the current).

h. A third non-linear effect is caused by *electrolysis*. This was shown by coil *D* which was watercooled and divided into four sections for this purpose. By doing so the primary coil, having a total impedance of 0.2 Ω, is shunted by the cooling circuit with a resistance of the order of 100 kΩ, which means that 10^{-6} of the main current is shunted off. Simple resistance measurements with a.c. show that in connection with polarisation etc. the resistance, at the frequency used, is a not very reliable quantity, changing with the voltage. Doubling the voltage and hence the current, as was done in *b*, the resistance can change considerably and hence an effect may arise of the order of our diamagnetic effect. Therefore we applied external oil cooling.

j. In connection with the *mutual capacity* the voltage across the primary coil causes currents through the secondaries. In the scheme (fig. 9) the mutual capacity C_m is placed at half the height of the coils. The impedance of the primary is Z_p, of the secondary Z_s, while the upper part of the compensator adds only minor contributions to this. The voltage of P is $\frac{1}{2}IZ_p$, so that the capacitive current is

$$I' = \tfrac{1}{2}IZ_p \cdot i\omega\, C_m.$$

The drop of voltage in the secondary circuit then becomes

$$I' \left(1\tfrac{1}{2}Z_s + \tfrac{1}{2}Z_s\right) = 2I'Z_s.$$

In this calculation, however, we neglected a second effect, viz. the mutual inductance. Now it can easily be derived that scheme b (fig. 10) is equivalent to a as concerns the potentials of and currents

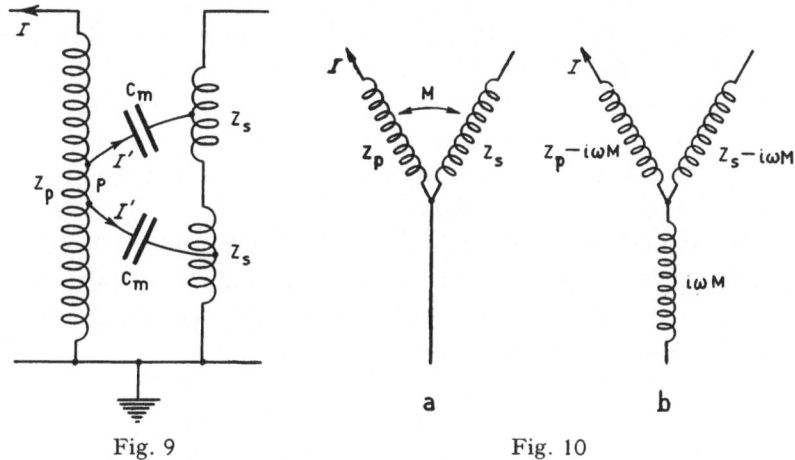

Fig. 9   Fig. 10

through the terminals. Hence the effect becomes

$$V_{mut\,cap} = i\omega\, I(Z_p - i\omega\, \Delta M)\, (Z_s - \tfrac{1}{2}i\omega\, \Delta M)\, C_m.$$

In our case $M < L_s$, furthermore the two coils oppose each other so that $\Delta M < M$ and these terms can be neglected. The term $i\omega\, I \cdot Z_p \cdot \tfrac{1}{2}\, i\omega\, \Delta M \cdot C_m$ corresponds to the expression for the self-capacity, but is much smaller.

Hence the main effect is

$$V_{mut\,cap} = i\omega\, I \cdot C_m\, Z_p Z_s, \tag{12}$$

which has to be compared with V_0 (see (1)).

Therefore the following ratio ρ is of interest:

$$\rho = \frac{C_m Z_p Z_s}{-M} = C_m \frac{\omega^2 L_p L_s}{M}\left(1 + \frac{R_p}{i\omega L_p}\right)\left(1 + \frac{R_s}{i\omega L_s}\right). \tag{13}$$

As table II shows, $R/\omega L$ is rather constant for the several coils (apart from D) and of course it is advantageous to let neither R nor

ωL become too large. The formulae for L_p and M are very simple (see (2)); the self-inductance of a rather flat coil (L_s) can be estimated with the model of a solenoïd ($\sim N_s^2 S_s/l_s$; S: surface, l: length) or of a flat coil ($\sim N_s^2 r_s$; r_s: radius). In 4 we shall consider C_m more in detail, applying the formula for a cylinder and a sphere. In both of these cases (13) can be much simplified, applying either the former or the latter formula for L_s. We obtain

$$\rho \sim \nu^2 \, N_p N_s S_p \cdot \frac{r_p}{r_p - r_s} \text{ or } \cdot \frac{1}{\log r_p/r_s}. \tag{14}$$

As r_s is determined by the quantity of the sample, r_p is determined by this formulae ($S_p \sim r_p^2$). The ratio is minimal if $r_p/r_s = 1.5$ and 1.6^5 respectively. More precisely the denominators in (14) contain the distance of the outer layers of the coils. In coil A we applied a factor of about 1.5, in later coils about 3. The ratio of the average radiï of the coils should be about 2.2.

A second effect that directly appears is the dependence upon N_p and N_s. As we started with a relatively large number of turns, this was strongly reduced as far as other effects allowed this. Furthermore the square of the frequency occurs in (14).

TABLE II

	A	B	C	D
$R_p/\omega L_p$ . . .	0.5	0.8	1.0	0.3
$R^s/\omega l_s$	1.6	0.7	0.8	4
$\dfrac{r_p^3}{r_p - r_s} \cdot 10^3$ . .	1.5	16	2.0	3.0
$N_p N_s \cdot 10^{-6}$ . .	17	8	0.3	0.12
$\left\| Z_p Z_s \right\| \dfrac{C_m}{M}$ . .	2 $\cdot 10^{-4}$	$0.6 \cdot 10^{-3}$	$0.6 \cdot 10^{-3}$	3 $\cdot 10^{-6}$

In the last line of table II the values ρ are given, calculated with the precise values of the several quantities and using the absolute value of $Z_p Z_s$ as a measure. As is clear, the decrease of this quantity is largely due to $N_p N_s$. Furthermore the product of the third and fourth line runs roughly parallel with the precise values of ρ so that simplification (14) is a fairly good one if we compare coils of similar shape. In coil D the effect equals the diamagnetic effect (see table V).

k. In the beginning (coil A) we used quartz tubes as carriers for our coils. Some weeks after the mounting of the apparatus *dielectric losses*, apart from normal capacitive effects, appeared. In connection with the occurrence in the formulae of the combination $i\omega$, the voltage in phase with the primary current contains the even, and that 90° out of phase the odd powers of ω. If losses are present, they can be described by a complex dielectric constant and then of course also other terms in the power series of ω appear, which is proved by the examination of the dependence upon the frequency. Drying the quartz coil in high vacuum, the coefficient of ω in M again could be decreased by a factor 20. Therefore later on we used pertinax.

Our sample container, which is made of quartz because of the easy cleaning, shows the same effects. A single touch of the fingers can cause a difference of 0.1 of the diamagnetic effect. In coil B the total effect even amounted up to twice this effect. A tin foil placed between the coil and the container prevents this quantitatively. Working with pincers and drying with phosphorus pentoxyde this trouble can be reduced. Furthermore on shielding, the dielectric properties of the substance cannot affect the measurements, changing the capacity of the coils. Of course we must prevent too large eddy current losses.

The capacitive effects and dielectric losses are the main reason for the dependence of the voltage upon the frequency. But in connection with the occurrence of quadratic terms it is always possible to find an extremum for one of the voltages by adding some parallel capacity.

4. Investigation of the self- and mutual capacity

Self-capacity

We examined the self-capacity somewhat more in detail as the effects connected with it have a strong influence upon the precision of measurement. The effective self-capacity can be found by means of natural frequency measurements, varying a condensor placed in parallel to the coil. If we plot $1/\nu^2$ against C ($1/\nu = 2\pi \sqrt{LC}$, see fig. 11), we get a fairly straight line, ranging from a parallel capacity zero up to some hundred times the self-capacity. This is strange as in the neighbourhood of the natural frequency the voltage distribution changes in the coil. The slope of the curve fits with the calculated L. In table III the experimental values of the self-capacity for the several coils are given.

In the calculation we shall first estimate the effect of the capacity between the layers. Calculating the impedance of the system in fig. 12, it follows that the capacity C_x placed across the part Z_x of the impedance Z, can be replaced by C_s if

$$C_s = \left(\frac{Z_x}{Z}\right)^2 C_x. \qquad (15)$$

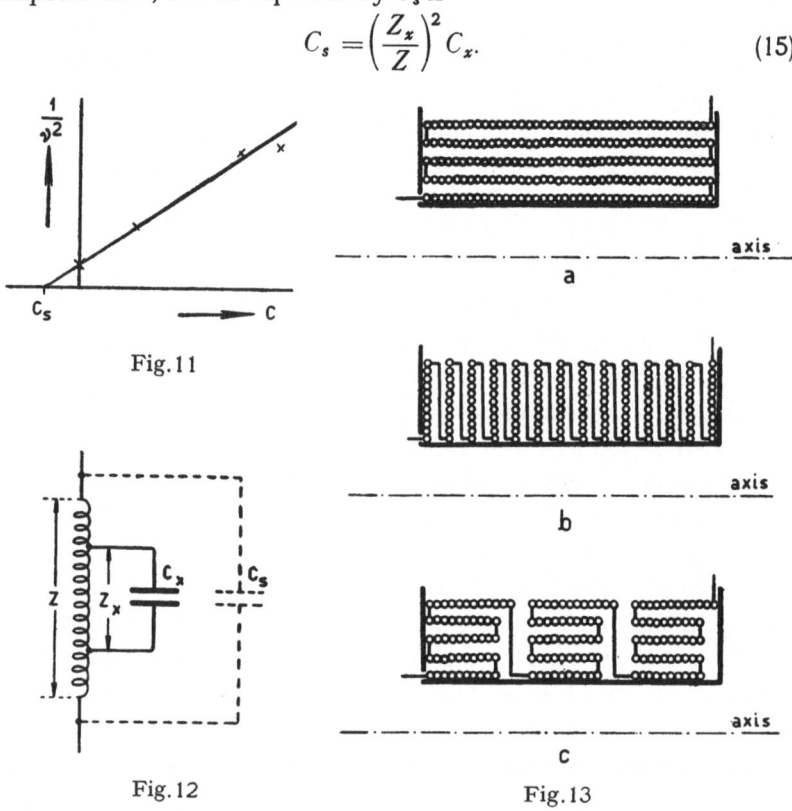

Fig.11

Fig.12

Fig.13

The second power can easily be understood. The voltage across C increases with the size of Z and besides the capacitive current flows through a larger impedance. Hence keeping the same total number of turns, the effect of capacitance decreases strongly if we divide the coil in several sections. Then the voltage between the neighbouring layers is lower and the impedance between them also (fig. 13). System b is favourable in this respect, but c can be constructed more easily.

If we have a capacity C_0 between two insulated layers, then it easily follows from (15) that the effective capacity between two connected layers, placed between the halves of the layers, is

$$C = \frac{4}{3} C_0.$$

This C_0 can be calculated with the model of a flat condensor in which the distance d between the layers gives the main error. In connection with the circular cross-section (see fig. 14), we add 10^{-2} mm to the thickness of insulation ($\varepsilon_r \approx 2$).

$$C = \frac{4}{3} \varepsilon \frac{O}{d} = 145 \frac{l_s r_s}{d} \text{ pF}. \tag{16}$$

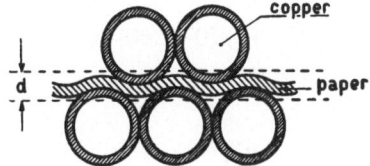

Fig. 14

In table III these values are given. If we have n layers, $n—1$ capacities are present, each having an effect C/n^2 according to (15) so that the contribution to the self-capacity is C/n.

TABLE III

Distances in mm Capacities in pF	$10^2 d$	l_s	$2 r_s$	n	$C = 145 \frac{lr}{d}$	C/n	C_s estim.	C_s meas.
A	3	50	30	20	3500	175	500	550
B	8	50	70	40	3000	75	330	300
C	8	6	25	20	130	6^5	19	25
D	9^5	3^5	28	14	75	5^5	13	18

There are several reasons that this value is much lower than the experimental ones. First of all both in this effect and in the mutual capacitance it is found that the capacity between the layers of coils, perhaps by reason of imperfect shielding, is larger than between solid plates. Between two single layers (coil B) we find experimentally a C_0 of 2600 pF, between a single layer and a group of layers 3100 pF, between two groups 3700 pF. Per plate we get an increase by a factor 1.2 (the diameter of the wire is 0.1 mm, the thickness of the insulation $2 \cdot 0.01$ mm). Secondly there is a large effect outside the coil between the first and the last layer and also between the sides and the end-layers. As is obvious, all these terms depend upon the distance between the layers and hence spacing of the turns by means of paper sheets gives improvement. Furthermore the coils become more regularly wound and compensate each other better. The capacities have been measured following the method described below under the mutual capacity.

In fig. 15 the several effects of coil B are indicated. The 3700 pF between the layers add 90 pF (calculated in table III: 75 pF). Between the outer layers we find 45 pF (in air). The side itself adds 60 pF, contributing to C_s according to (15) 60/4 = 15 pF. Assuming a mean $\varepsilon_r = 2$, we find a sum total of

$$(90 + [45 + \tfrac{1}{4} (100 + 140 + 60)] \, \varepsilon_r = 330 \pm 80 \text{ pF}$$

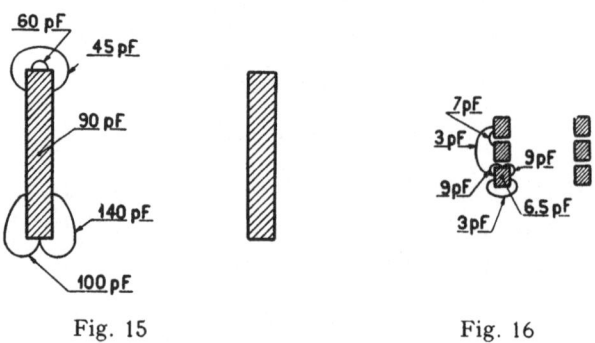

Fig. 15   Fig. 16

In fig. 16 coil C is drawn with the additional capacities; it is drawn on the same scale as coil B in fig. 15. Placing the coil in a medium with $\varepsilon_r = 2$, a total capacity of 19 $\pm$ 5 pF follows. By considerations of similarity the other coils can be calculated. Table III shows that the order of magnitude fits now.

As a result we can say that small coils are favourable, paper sheets and winding in sections necessary. In the two last coils the capacity has become entirely distributed. It will perhaps be somewhat more than has been taken into account, e.g. in the supply wire.

Mutual capacity

The mutual capacity can be measured according to (12). Changing Z_p or Z_s, a change in the voltage is obtained. In table IV the values are recorded as C_m meas.

For the calculation of the capacitive effect between one primary and one secondary layer we shall compare two models. We apply the formula for an infinitely long cylinder, calculated over the length of the secondary coil:

$$C_{cyl} = \frac{110 \, \varepsilon_r \, l_s}{4.6 \, {}^{10}\log r_p/r_s} \text{ pF}$$

as a lower value and that for a spherical condensor with radii equal to those of the primary and secondary coil:

$$C_{sph} = 110 \frac{\varepsilon_r r_p r_s}{r_p - r_s} \text{ pF}$$

as an upper value (table IV: $\varepsilon_r = 2$). These numbers differ of course. We assume a mean value C_0, giving the cylindrical case twice the weight.

TABLE IV

Distances in mm Capacities in pF	$2 r_p$	$2 r_s$	l_p	l_s	C_{cyl}	C_{sph}	$C_0 = \frac{2 C_{cyl} + C_{sph}}{3}$	n prim.	k	C_m estim.	C_m meas.
A	45	30	200	50	13	10	12	6	10	120	200
B	210	70	360	50	5	11	7	1	8	56	100
C	70	25	280	20	2.2	4	3	1	6	18	20
D	90	28	360	17	1.6	4.5	2.5	1	6	15	12

For a further investigation we use coil B. It appears that the capacity of a several layer coil is larger than that of one consisting of flat plates. An increase of 1.2 for the secondary and 1.5 for the primary has been found. Hence for coil A, containing more primary layers, there results a factor 1.8.

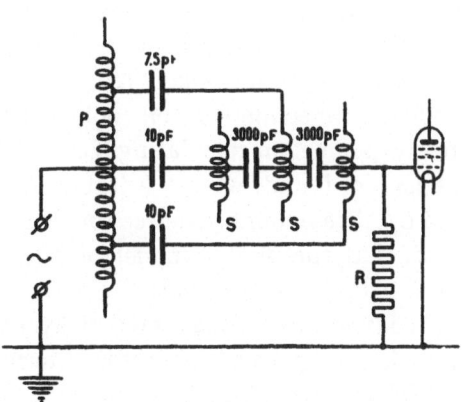

Fig.17

The effects of the several parts of the secondary coil can be measured fairly well by applying a drop of voltage between the primary coil and the earth. The secondary layer under examination (fig. 17) is earthed over e.g. $R = 10 \text{ k}\Omega$ so that the voltage across this resistance is a measure for the unknown capacity. The other secondary layers are coupled by 3000 pF with this layer and hence are also measured $(C_m \sim 10 \text{ pF} < 3000 \text{ pF})$. If we wish to eliminate this effect, we earth or shield these layers, which does not disturb the measurement as the impedance of 3000 pF still is much bigger than the 10 kΩ ($\omega = 2000$).

Earthing the several layers of the coil, a fair additivity of the capacity is found. It follows that the effect of the inner layer r (fig. 18) equals that of the outer layer p, the side q causing $\frac{3}{4}$ of this effect.

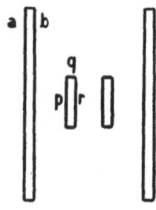

Fig. 18

The supply wire with a radius 10^{-3} of the secondary coil, but with length 5 times as great, gives a similar effect. This means that the effect of the secondary coil is larger by a factor $3\frac{1}{2}$ compared with the outer side of one layer. By reason of symmetry it follows that the outer side a of the primary coil will also add to the capacity; experimentally a factor 1.8 for an open primary (B), 1.5 to 1.2 for a more closed one is found. In connection with oil cooling ε_r will be 2 in the mean, as was already assumed in calculating C_0.

Combining the factors found, we get k (table IV) and thus the estimated value of C_m. These numbers agree roughly. Therefore it is of interest to recall that the simple model of the cylinder gives results, one order of magnitude too small.

Small coils will be favourable; using a several layer primary coil it is advantageous to let the current enter the intermediate layers, which are shielded so that the outer layers will have the lowest voltage. In connection with eddy currents and the increase of the self-capacity we do not apply tin foils as shield between the primary and secondary. The application of one earthed outer layer may give some improvement, but then the coil is coupled with several thousand pF to the earth.

5. Design of coils

A. In the first arrangement we used one main and two half secondary coils, one above and one below the central coil. In this manner stray fields are eliminated rather well while furthermore displacements in the primary coil will have less effect. Here the substance can be placed only in the main coil or outside of it so that the effect is not doubled as in later constructions; in table V, where data concerning the coils *A*, *B*, *C* and *D* are given, *f* has been halved; in one coil it was 0.4.

The coils have been wound upon quartz tubes having a low thermal expansion. To obtain sufficient precision a rather large number of turns was used. The natural frequency of the main coil was 6500 Hz while on adding a parallel capacity the *LC* of the two half coils became equal to that of the main coil. Therefore the term with ω^2 in *M* became very small. Nevertheless experiments showed a strong dependence upon the frequency. Some weeks after the mounting of the apparatus the coëfficient of ω in *M* increased by a factor 20. After drying in vacuum with a coal tube in liquid air the original properties could be recovered, clearly demonstrating the dielectric losses caused by water.

The dependence upon the frequency was 0.01 to 0.1 of our effect per Hz; thermal effects of the secondary coil seemed disturbing too The precision of measurement became some tenths of our effect.

B. In this construction we tried to decrease the dielectric losses by winding the coil upon pertinax. To prevent thermal effects we placed the secondary coil in a D e w a r flask and also put a small D e w a r flask inside this coil. This led to an increase of the diameters of the coils and hence to a decrease of the filling-factor ($f = 0.1$). The dielectric losses due to the glass are reduced by earthing the silver-coating. Unfortunately the latter introduces eddy currents, but for the reduction of dielectric effects of the carrier it is of great importance.

Two equal secondary coils, each carrying a large number of turns, are used. At first they were rather roughly wound and then badly compensated each other. It appeared that *L* was 10 and 11 H, C_s 1400 and 850 pF. After rewinding the coils with papersheets (0.05 mm) between the layers, this became 12 H and 300 pF while the coils differed only slightly. The natural frequency became 2700 Hz.

The voltage changed about 0.2 of the effect per Hz and a number of higher harmonics remained after compensation. A rise of the temperature of the primary coil caused a large effect. The precision was of the order 50 pro mille of the diamagnetic effect.

C. When constructing a third apparatus a decrease of the capacitive effects seemed to be necessary. The number of turns was strongly reduced while the secondary coils were wound in three sections, seperated by 1 mm thick walls and with papersheets (0.05 mm) between the layers. After winding, the whole space around the secondary coils was filled with melted wax in vacuum to reduce dielectric losses. The natural frequency was found to be 130 kHz while the self-capacity was 25 pF and the mutual capacity 20 pF. We used a one layer primary coil with a second layer at the ends with a length one third of the diameter of the coil. Thus the homogeneity became 10^{-5} per mm over one third of the length of the coil.

The primary was cooled by means of a swift oil stream. As the dissipated energy increased strongly, we also cooled the space between the secondary and primary. The precision of measurement became some pro mille.

Unfortunately the energy required was not available. As the resistance of the primary coil had become somewhat high, we took a very big wire. A supply pipe through which water streamed was laid three times around a transformer and shortcircuited at another point. The primary coil itself was something between a helix and a H e l m h o l t z coil. The same secondary as previously was used. The resistance of the primary was about $3 \cdot 10^{-3}$ Ω, the current could be 500 A. Potential wires for the compensator were soldered at a short distance upon a rectilinear piece of the pipe. The experiment showed that it was difficult to find a minimum at the same place for the mutual inductance and the eddy current losses. The cross-section of the lead pipe was too big.

D. In the last construction the primary coil consists of 25 m capillary, with an inner diameter of 2.4 and an outer diameter of 3.8 mm while the turns are laid at a distance of 4.2 mm. Six extra turns have been placed at each end to obtain a homogeneous field. To begin with, the coil was divided in 4 sections to allow 1 dm^3 water to pass per minute. Afterwards in connection with electrolytical effects the whole coil was placed in a box and externally cooled with oil. In these oil cooled constructions, consisting of pertinax etc., the closing sometimes gives difficulties. We placed the pertinax tubes in narrow grooves, with rings of rubber, in a lower and an upper plate, which were forced together with some bolts (fig.1). At some minor points, e.g. the places where the supply wires of the secondary coil are carried through, material as piceïne can be used.

The secondary coil (fig. 19) consists of 4 sections, 350 turns each, at one mm distance; the inner diameter is 25 mm, the outer diameter 31 mm, the length 17 mm. The distance of the secondary coils, intended to be 70 mm, was found in the measurement with the sample to be 69 mm. The surfaces of the coils differed 7 pro mille.

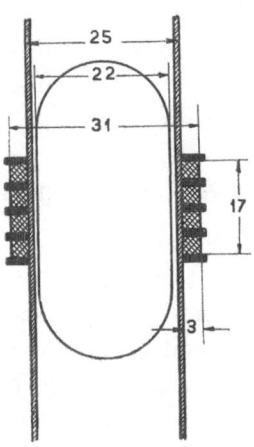

The eddy currents were not entirely compensated together with the mutual inductance. With a small piece of zinc (M, 2g), placed between the primary and secondary coil, this could be attained. The dielectric effects of the carrier were eliminated by means of an earthed tin foil, placed inside the secondary coil.

Fig. 19

With this last construction enough precision could be obtained while the energy required could be furnished by an energy amplifier and carried off by means of a cooling-system. All the measurements were performed with this apparatus.

TABLE V

	unit	A		B		C		D	
		p	s	p	s	p	s	p	s
$2\,r_{thread}$	mm	0.6	0.1	0.6	0.1	1.0	0.18	4.2	0.1
N		2000	8700	600	13,500	320	900	88	1400
R	Ω	17	1750	30	7,000	1.4^5	45	0.06	260
L	10^{-3} H	35	1100	40	10,000	1.4^5	60	0.2	65
volume sample	10^{-6} m³	16		30		6		9	
f			0.2		0.1		0.4		0.4^5
M	10^{-3} H	80		100		0.5		0.2^5	
ωMI	V	25		50		12		11	
I	A		0.3		0.5	24		45	
nI	10^3 A/m		3.0		0.8	24		11	
$I^2 R$	W		1.5		7.5	800		120	
$V_{self\,cap}$	diamagnetic	1000		7000		0.9		1.2	
$V_{mut\,cap}$	effect	180		1000		2.5		1.0	

Finally we shall suggest some improvements of the apparatus used. As a whole a smaller apparatus is possible while the same amount of power is sufficient. The length of the secondary could be 15 mm, the diameter 22 mm, the number of turns 1400, the diameter of the thread 0.1 mm. The volume of the substance then becomes $4 \cdot 10^3$ mm^3. The length of the primary would be 200 mm, the diameter 50 mm, the number of turns 50, the diameter of the stranded wire 4 mm. The current then is about 80 A while the effect of capacitance decreases.

6. Auxiliary apparatus

Cooling and thermostat system

A temperature constancy of about 10^{-3} °C during some seconds proves to be necessary for the secondary and primary coil. As the last coil dissipates energy, we use two separate cooling systems; in each of them a cogwheel pump presses spindel oil through the apparatus. A large watercooled basin with a swift stirrer provides for a constant temperature, about 20°C. Half an hour after the beginning enough constancy is obtained. In connection with the fact that the cogwheel pump does not work quite regularly the oil stream gives a varying pressure upon the secondary. This is smoothed by a factor 10 by means of a closed air reservoir, placed upon the pressure side.

Hoisting apparatus and relais

The secondary coils are placed above each other so that the cuvette, suspended on a thin quartz rod, can move up and down between the coils with a hoisting apparatus.

The current for the d.c. engine, used for hoisting, can be interrupted with a relais P (fig. 20) which also operates the relais cutting the supply wire of the galvanometer. This latter relais is critically adjusted by means of its supply current so that it takes some time before it works and switching effects don't disturb the galvanometer. Another relais Q commutates the engine current and operates the relais for the compensator resistances. As a rule we apply telephone relais; only Q is different, somewhat slower.

In fig. 20 the movable parts have been drawn in the case of non-excited relais. When the relais work, they are attracted by the coils. If Sa (sample) meets A, the relais P is exited, the engine current interrupted. Now Q also carries current and commutates, while q is closed. If the coil of P is shortcircuited (switch S) for a moment, after a measurement has been done, Sa leaves A and p is closed. Then Q gets a current through p and q, since the last contact stayed for a short time at its place. If Sa reaches B, the engine stops, p drops out and Q commutates.

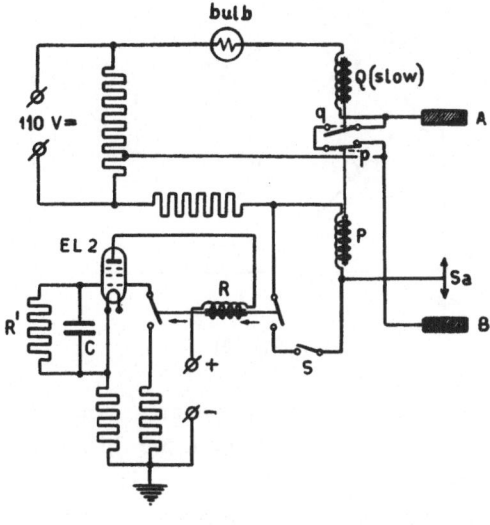

Fig. 20

The switch S is supervised by the observer and also by an $R'C$-element which, placed in the grid circuit of an EL 2, switches a relais R. Then the condensor gets the negative voltage across the cathode resistance ($2k\Omega$) and the anode current again is reduced. The $R'C$-time is 5 seconds.

Low frequency generator

Our demands of a constant frequency and a small contribution of higher harmonics are fulfilled by an RC-generator. The system (fig. 21) consists of a W h e a t s t o n e bridge in which the galvano-meter is replaced by the first valve of an amplifier while the second valve furnishes the current for the bridge after a phase turning of 360° (Z a i s e r, 1942). The branches of the bridge consist of an RC-element in parallel, one in series, a glow-lamp and an ohmic resistance. Though the part of the circuit, determining the frequency, has a bad selectivity, still the distortion is low. In connection with back-feeding the phase is of great interest and changes rapidly·in the neighbourhood of the generated frequency $\omega_0 RC = 1$. Therefore a small change of the frequency already causes a suppression of the occurring signal.

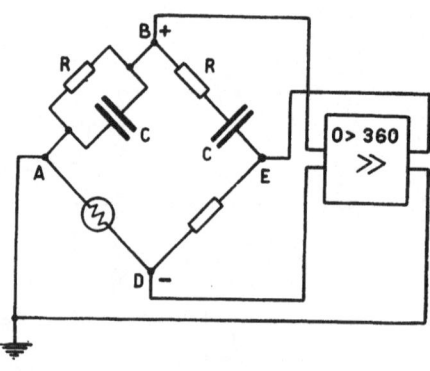

Fig. 21

If $\omega = \omega_0$, we have for the impedances of the RC-elements: $Z_{ser} = 2Z_{par}$, independent of ω_0. The other branches of the bridge provide for the stabilisation of the voltage. They contain a glowlamp of which the non-linearity stabilises, but gives only a low distortion. The amplitude of the sine wave, established at less than half the

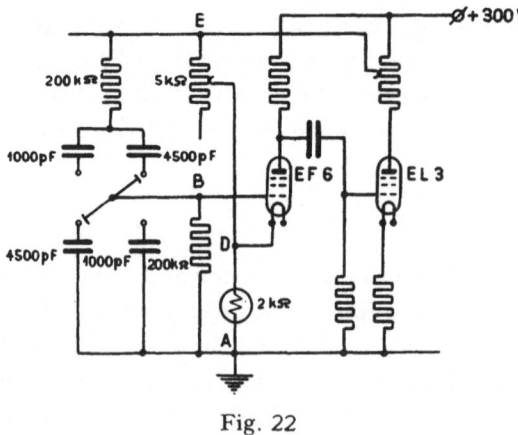

Fig. 22

maximum value, is roughly adjusted by means of the ohmic resistance, which then has half the resistance of the glowlamp (see fig. 22).

Energy amplifier

The amplifier (Multiper, The Hague), getting an input voltage from the generator, furnishes 200 W energy. In connection with non-linearities and capacitive effects a constant output of low distortion is of interest. The effects of the non-linearities are decreased by applying iron wire lamps (no. 329; 10/30 V, 1.15 A) with which we gain a factor 5 in constancy while about a quarter of the available energy is dissipated (see fig. 2). For a low distorsion a good matching is of interest. Using a transformer (10 : 1) the coil properties become such that with a suitable condensor ($C' \approx 5$ µF) the wattless power can be carried away. LC in parallel seems to be better for the energy amplifier. With a second transformer (1 : 3) L/RC is matched to the amplifier and stabiliser.

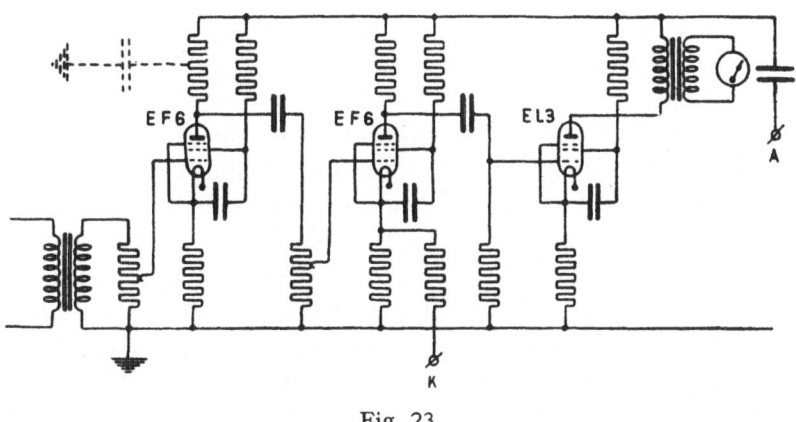

Fig. 23

Voltage amplifier

A three valve amplifier (fig. 23) with an input and output transformer is used. To obtain magnetic shielding the input transformer has been placed in an iron box with a thickness of 8 mm. The amplifier has been put in an aluminium box with plates between the tubes to avoid generating, moreover all anode voltages are separated by means of RC-elements.

Two EF6 and one EL3 tubes are resistance-coupled while the anode of the EL3 is coupled with the cathode of the second EF6 by means of a filter. The amplification is more than sufficient.

Filter

The secondary coil is tuned with a suitable condensor. In the amplifier we apply an *RC*-filter (B o u r g o n j o n, 1943). In fig. 24

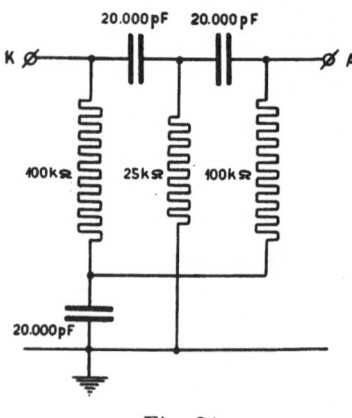

Fig. 24

the system is reproduced, the values apply for the case $v = 100$ Hz. The principle of the system is backfeeding through an element which depends upon the frequency. This gives a relation both in phase and amplitude so that a high selectivity is obtained. The sharpness of resonance is represented by $Q = f\alpha/4$, in which f is the amplification and α the match-factor in the scheme. In the arrangement a $v/\Delta v_{\frac{1}{2}} = 200$ can be reached so that the amplitude is halved at about 2 Hz. In the peak of the "resonance curve" the phase changes with 90° while the amplitude changes only $\frac{1}{2}\sqrt{2}$; this provides for a beautiful adjustment of the phase.

Galvanometer

We apply a d.c. galvanometer equipped with field coils that can be supplied with alternating current. The galvanometer gets a deflection if a voltage of the same frequency is put on the measuring coil and a phase relation is fulfilled (Mi l a t z, 1937). The resonance curve is sharp, the half value breadth $v/\Delta v_{\frac{1}{2}} = 60$. The sensitivity of the third harmonic is not described by this number as the current exciting the galvanometer also contains the third harmonic; it amounts up to 10^{-2} of the fundamental.

We use M o l l galvanometers, one of an older type with field coils and a more modern one with a permanent magnet which had to be provided with extra coils. Though the μ_r of the magnet steel is not very high, the sensitivity is similar if the two coils (2 · 250 turns) are pushed close to the slit. For the ordinary damping one of the coils of the first galvanometer carries direct current.

Of great interest of course is the adjustment of the phase. To obtain this, we adjust for a maximum current with condensors in series with the field coils; this means resonance (fig. 25). Then the capacity

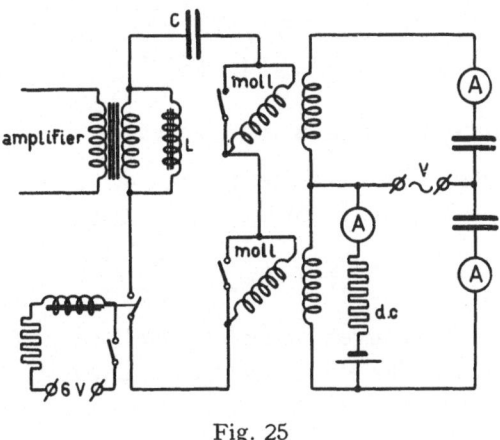

Fig. 25

across one galvanometer is increased and across the other decreased until the current has become $\frac{1}{2}\sqrt{2}$ of its original value and both galvanometers have become sensitive for voltages 90° out of phase. A further adjustment is reached by applying the phase change of the RC-filter in the voltage amplifier for the case of resonance. We adjust in such a way that the galvanometer is made insensitive for a voltage 90° out of phase, as is furnished by the compensator.

By doing so the first M o l l galvanometer can be put in phase with the mutual inductance, the other in phase with the losses. The linear scale provides for swift working as only the amplitude needs to be read off.

REFERENCES

B o u r g o n j o n, L. R. (1943). T. Ned. Rad. Gen. **10**, 209.
M i l a t z, J. M. W. (1937). Thesis Utrecht.
M o u l l i n, E. B. (1938). Spontaneous fluctuations of voltage.
Z a i s e r, W. (1942). El. Nachr. Techn. **19**, 228.

5

CHAPTER V

INDUCTANCE APPARATUS WITH VARYING COEFFICIENT OF INDUCTION

1. Discussion of the measuring-generator

In this chapter we shall consider the measuring-generator in greater detail. According to II (15) the following holds:

$$V = -f\varkappa \cdot i\omega \frac{\Phi_0}{2\sqrt{2}}, \qquad (1)$$

in which $2\sqrt{2}$ is the conversion factor from the double amplitude to the effective value. The filling-factor, according to II (13), is defined by

$$\Delta\Phi = f\varkappa \Phi_0, \qquad (2)$$

where $\Delta\Phi$ is the total change in the flux threading the coil when we bring the sample into the field. f can be estimated by considering the two idealised cases already mentioned in the introduction in II.3.

One magnet and one sample

a. If we bring a substance into a *homogeneous field*, the number of lines of force threading a secondary coil will change (fig. 1). If the substance has a cylindrical shape, according to IV (7)

$$f_{cyl} = \frac{S_r}{S_s}.$$

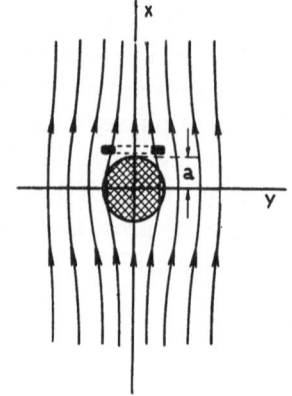

Fig. 1

The formula can be applied if a rather flat coil crosses the cylinder of magnetic material through a narrow slit. In this case we can take the diameter of the cylinder equal to that of the coil (see II fig. 17) so that $f = 1$. This, however, is an upper limit.

If the substance has a spherical form, according to IV (6)

$$f_{sph} = \frac{v_{sph}\, K_s}{S_s\, N}.$$

This case is an unfavourable one and will give a lower value. If we use a flat coil with N turns, which passes the moving sphere at a distance $x = 1.2\, a$ from the centre of the sphere (see fig. 1 and IV fig. 4), there follows with the formulae in IV.1:

$$f = \tfrac{2}{3}\, \frac{a^3}{(\rho^2 + x^2)^{3/2}}.$$

Substituting $\rho = a$, f becomes 0.2.

In an intermediate case f might be 0.4.

$b.$ Next we shall employ an *iron toroid* with a slit through which our sample moves (see fig. 2). According to I (1) and I (4) the following holds:

$$\sum_i H_i l_i = nI \tag{3}$$

and

$$\varphi = B_i S_i,$$

where l is the length, taken along the iron circuit, while i indicates the several parts of the circuit: sample (1), air (2) and iron (3). S_i is the surface of one of the parts which is assumed to be constant over this part; n is the number of turns of the primary coil. If we apply a

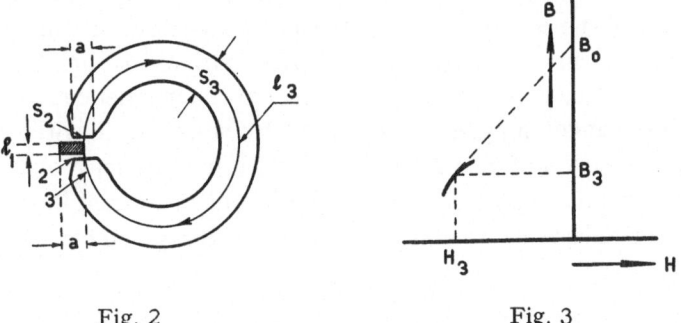

Fig. 2   Fig. 3

permanent magnet, its BH-curve is of importance (fig. 3). As the changes under consideration are small, we can represent the changes B_3 by the following formula:

$$B_3 = B_0 + \mu_0\, (\mu_r)_3\, H_3,$$

where B_0 is the point of intersection of this line with the B-axis, μ_r the differential permeability. A simple calculation yields:

$$\Phi = \frac{\mu_0\,NnI + N(B_0l/\mu_r)_3}{\sum\limits_i (l/\mu_r S)_i},\tag{4}$$

in which N is the number of turns of the secondary coil. As is obvious on applying an electro- or permanent magnet, either the former or the latter term in the numerator is applicable.

In (4) the properties of the substance enter only throug $(\mu_r)_1$ in the denominator. For the calculation of f we have to consider the difference of Φ with and without substance. Then we obtain:

$$\Delta\Phi = \frac{\mu_0\,NnI + N(B_0l/\mu_r)_3}{\sum\limits_i (l/\mu_r S)_i} \cdot \frac{(l/\mu_r S)_1}{\sum\limits_i (l/\mu_r S)_i} \cdot \left(\frac{\varkappa}{\mu_r}\right)_1 = \Phi \cdot f \cdot \varkappa.\tag{5}$$

Sometimes $l/\mu_r S = r$ is called the magnetic resistance of the piece of matter. If the resistance of our sample is r_1, ot the other part of the circuit r_0, we get:

$$\Delta\Phi = \frac{Cr_1}{(r_1 + r_0)^2},$$

which yields a maximum if $r_1 = r_0$ so that [1]

$$f = \frac{(l/\mu_r S)_1}{\sum\limits_i (l/\mu_r S)_i} = 0.5.$$

It must be noticed, however, that in the case of a permanent magnet the linear relation between B_3 and H_3 is only fulfilled in a small part of the BH-curve, while the precise position of this point upon the BH-curve is determined by the width of the slit. We shall not consider this point further. The effect of the magnetic resistance of the permanent magnet is decreased by shunting it with other poles (see below and also 3, apparatus with moving magnets).

If we apply an electromagnet, the coefficient of mutual induction is of importance as it can easily be measured. It is given by

$$M = \frac{\nu_0\,nN}{\sum\limits_i (l/\mu_r S)_i}.\tag{6}$$

E.g. in the trial apparatus (IV.3) it was observed that the mutual inductance depended upon the magnet current, decreasing by a factor two upon increasing the current. Also if we apply a permanent

---

[1] This definition of f corresponds to $H_1l_1 = f \cdot nI$.

magnet, the introduction of some additional turns will be favourable for the knowledge of the denominator in (5). In this case the denominator can be decreased if $(u_r)_3$ is large, which is the case in steeper parts of the magnetisation curve.

Recapitulating the cases a and b, we can expect that if the secondary coil is placed close to the slit, f will be about 0.5.

Several magnetic poles or samples

For an increase of the effect and improvement of the sinusoïdal form of the voltage it is of importance to use more than one magnetic pole or sample. If p is the number of poles, q the number of samples and r the number of revolutions of the apparatus, the frequency becomes

$$\omega = 2\pi \, pqr$$

and hence

$$\omega\Phi_0 = 2\pi \, pqr \, \Phi_0 = 2\pi \, r \, \frac{v_{tot}}{v_0} \, \Phi_{tot},$$

where Φ_{tot} is the total flux threading the air slits; v_{tot} is the total volume of the substance, v_0 that of one single sample. Obviously

$$\frac{j}{v_0} = \frac{1}{v_{magn}},$$

where v_{magn} is the effective volume of the air gap and iron circuit. Hence

$$V_{eff} = \frac{\pi}{\sqrt{2}} \chi \cdot r \, v_{tot} \cdot \frac{\Phi_{tot}}{v_{magn}}. \tag{7}$$

We shall compare r with the number of revolutions of a three-phase motor, being about 25 per second, so that $r = 25 \, \alpha$. Furthermore we assume an average diamagnetic susceptibility of $\varkappa = 0.7 \cdot 10^{-5}$. Then

$$V_{eff} = 0.4 \, \alpha \cdot \frac{v_{tot}}{v_{magn}} \cdot \Phi_{tot} \text{ millivolts.} \tag{8}$$

For the application of this result it is of course necessary that the apparatus has been constructed well.

Applyirg a field of 0.15 Vsec/m² (1.5 kG) and a secondary coil of 5000 turns, laid around a surface of $5 \cdot 10^{-4}$ m², the total surface of the coil becomes 2.5 m² and $\Phi_{tot} = 0.4$ Vsec. We can use e.g. 10 cuvettes, $f = 0.5$ so that $v_{tot}/v_{magn} = 5$, and $\alpha = 1$ sec⁻¹. In that case $V_{eff} = 0.8$ mV.

Experiments point out that the disturbances are about $1 \cdot 10^{-5}$ V and hence that measurements with a precision of about 10 pro mille are possible ($\Delta x = 1 \cdot 10^{-6}$).

An increase of the effect can be obtained, according to (7), by an increase of the total quantity of the sample, the number of revolutions of the apparatus and the number of poles, apart from an increase of the numerator in (4). In (7) we took Φ_{tot} and v_{magn} together as in this ratio S can be cancelled, while l is more or less fixed by the condition for f, the important quantity v_{tot} thus becoming a proportionality factor in V_{eff}.

The distance of the impulses of course is of great importance. In fig. 4a an impulse is drawn. The axis of ordinates gives the flux threading the secondary coil, the abscissa the time, or as the body

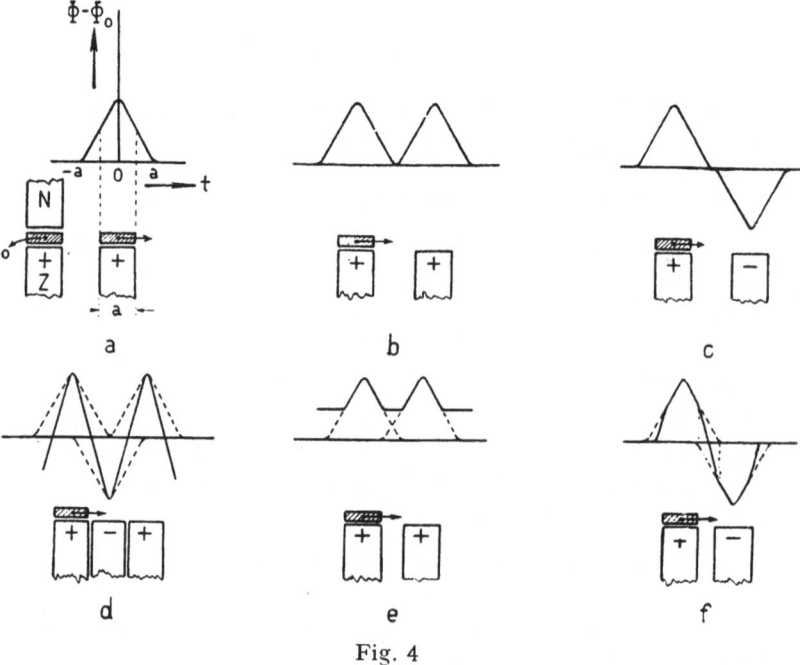

Fig. 4

moves at a uniform velocity, also the place of point O, the centre of the sample. N is the north pole of the magnet, always kept on the same side. The $+$ or $-$ sign indicates the direction of winding of the secondary coil.

Placing the poles or the samples at a distance equal to their size, we obtain curve *b*. In *c* the coils have been connected oppositely one after another. The flux change is doubled, the frequency is halved and hence the voltage stays constant, as may be seen at once on applying (7) or (8). In *d* the number of poles has been doubled; the amplitude is also doubled because Φ_{tot} has been doubled. The frequency stays the same as in case *b*. The system with oppositely connected coils has advantages both by reason of the improvement of the sinusoïdal form of the curve and the decrease of the variations in the main current. Furthermore a deviation in the distance of the impulses is not so dangerous; compare e.g. *e* and *f* in this respect.

2. Measurement of the voltage. Disturbances

The voltage across the secondary coil can be compensated with another voltage. This alternating voltage can easily be obtained by moving a piece of magnet steel, connected to the movable part of the apparatus, along a coil. In a second system direct current can be applied. A single turn laid around the sample and carrying a small current can compensate the effect of the substance so that the secondary coil has only to indicate a zero voltage. If we branch this current off from the magnet current, a null-method with respect to the current is obtained. In both methods changes in the frequency are allowed.

The connection of the moving parts and the indicating instrument is of importance as disturbing effects may arise from it. As far as could be seen, sliding-contacts worked well and the contribution of the fluctuations in the voltage seemed to be smaller than those of other origin. We also tried a transformer coupling (fig. 5) in which one coil and half the iron circuit moved while the other part remained at rest Due to remanent magnetisation of the iron a large null-effect occurred and the system was abandoned.

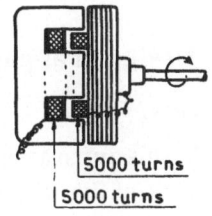

5000 turns

5000 turns

Fig. 5

Of the *disturbing voltage* of about $1 \cdot 10^{-5}$ V we observed the following contributions:

a. The largest effect seems to be due to the vibrations of the apparatus. If the construction is not sufficiently solid, the coils will vibrate in the magnetic fields and generate voltages. Possibly effects related to magnetostriction occur, which means that the vibrations should cause changes in the magnetisation of the iron.

b. Also the motion in the field of the earth gives effects. If the axis of the instrument is directed along the lines of force of this field, they are avoided. The field of the earth can also be compensated with a H e l m h o l t z coil.

c. The sliding-contacts seem to introduce only minor effects. Replacing the moving coil by a resistance, they could be examined to a certain extent.

d. If we apply an electromagnet, the current variations have effect. They can be reduced rather well with the system reproduced in fig. 4*c* or 4*d.*

e. Electrostatic effects can be eliminated by wrapping a shield around the coil.

f. The motion of pieces of metal with respect to the magnet has to be avoided.

Apart from these effects we wish to call to mind some factors that might reduce the voltage of the generator. There is a possibility for alternating voltages, upon shortcircuiting the primary coil by the battery which supplies the magnet current. Therefore one should place a coil with a large self-inductance in series with the primary coil so that a possible decrease by a factor two is prevented. As concerns the iron circuit it is of importance that the lines of force are concentrated in the air gap as much as possible and that the alternating flux arising there threads the secondary coil.

3. Design of apparatus

Trial apparatus

In this apparatus (fig. 6) we apply a small electromagnet; using a current of 2 A, the field B is about 0.25 Vsec/m². The circumference of a circular disc of wood, furnished with 16 holes which can be filled with the substance, moves through the air gap. Upon one of the pole pieces a secondary coil of 7000 turns has been placed. The primary coil consists of 3000 turns. The surface of the pole pieces is 500 mm², the distance 20 mm. Both the calculation with M, equal to 0.6 H on applying (6), and the experimental value of B yield a $\Phi = 1$ Vsec.

The disc is placed upon a 4 m long axis, coupled with a three-phase motor. The frequency is about 400 Hz; in (8) $\alpha = 1$ sec^{-1}. The total volume of the substance is $0.6 \cdot 10^{-4}$ m^3, the effective volume of the circuit $1.2 \cdot 10^{-5}$ m^3. Hence

$$V_{eff} = 0.4 \cdot 1 \cdot 5 \cdot 1 = 2\,\text{mV}.$$

Indeed the reproducibility proved to be better than 10 pro mille of the diamagnetic effect.

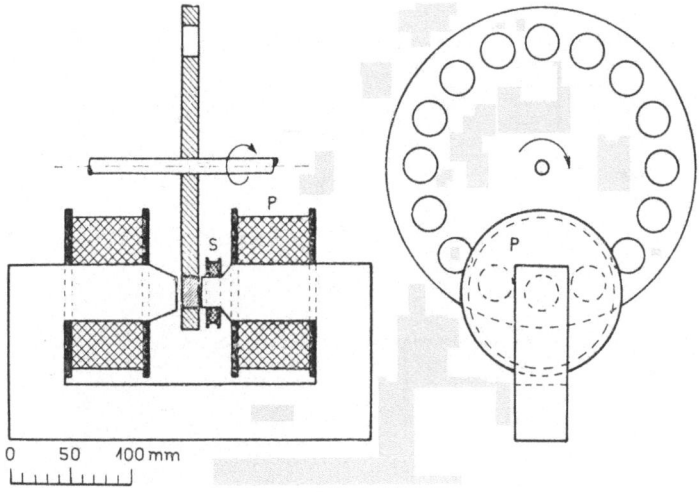

Fig. 6

Apparatus with moving magnets

We had a ring-shaped Ticonal magnet at our disposal (*a* in fig 7). It is mounted upon a brass axis (*b*), suspended in ball-bearings. Upon this axis also two systems of pole-pieces A en B have been placed; in fig. 7 they were drawn above each other. Each system consists of 8 pairs of pole-pieces (*c* and *c′*), constructed from transformer plate to prevent eddy current losses. Instead of separate coils for each of the 8 pole pieces, one coil (*d*) was used for the whole system (diameter thread 0.08 mm). By doing so the maximum diameter of the system can be kept small, the weight of the copper coil is less and finally the space between the pole pieces would not allow a large number of secondary turns. The thickness of the pole pieces is constant all over their length so that, placing them side by side upon the axis, they stand rather far from each other near the pole gap (*e*). By placing also a second system upon the axis, the apparatus becomes more symmetrical. It is also supplied with a secondary coil (*d'*), with opposite direction of winding. Thus each system threads a coil while the other one lies outside and we can expect an effect as is drawn in fig. 4*c* or 4*d*. Upon the axis the systems are coupled magnetically by means of an iron ring. As is obvious, there will be a strong leakage, especially inside the magnet ring through the axis. The lines *f* represent the cross-section through the box, keeping the system together.

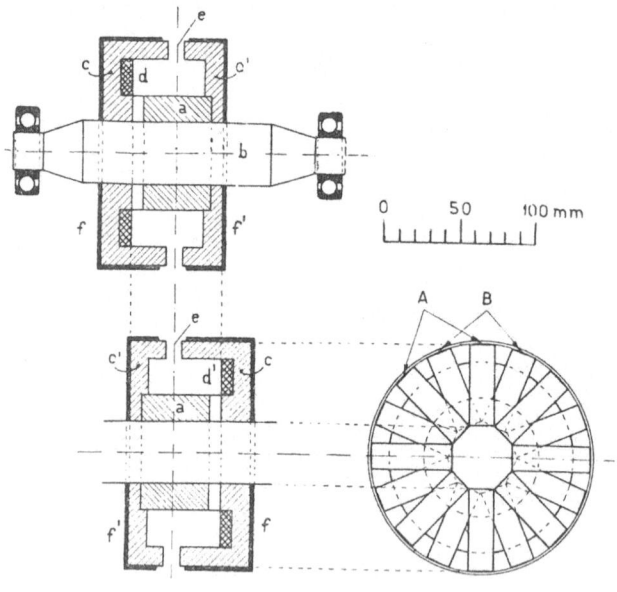

Fig. 7 [1])

If the sample comes into one of the pole gaps, a change in the flux arises, which goes together with a change in the flux both through the permanent magnet and through the other 15 pole pieces. As the effects of 14 of these fluxes oppose each other, the total effect will not be changed. But we have the advantage that the Ticonal magnet is shunted by the other poles so that the flux change will become larger by this reason.

For the size of the effect it is of importance that the optimum point of the *BH*-curve of the permanent steel is used. The length of the magnet steel is about 40 mm, of the air gap 11 mm; the surfaces are $2.8 \cdot 10^{-3}$ m² and $16 \cdot 200$ mm² $= 3.2 \cdot 10^{-3}$ m² respectively.

---

[1]) The lines *f'* have erroneously been drawn fat.

Applying (3) this seems to be compatible with $B_3 = 0.26$ Vsec/m²
and $H_3 = -5 \cdot 10^4$ A/m inside the magnet, while in the airgap we
can expect $B_2 = 0.23$ Vsec/m². Experiments show $B_2 = 0.05$ Vsec/m²
This means a leakage by a factor 4.5. The secondary coil consists of
4200 turns having a total surface of 13 m² so that $\Phi = 0.7$ Vsec.
Applying (8) and introducing $f = 0.4$, $\alpha = 1.4$, we get

$$V_{eff} = 0.4 \cdot 1.4 \cdot 0.4 \cdot 0.7 = 0.15 \text{ mV}.$$

It proved to be possible to observe this effect with a relative precision of 0.1. The form of the curve of the voltage appeared to be rather
good. The frequency was about 250 Hz.

We examined the form of the voltage also by placing an a.c. carrying turn in the pole gap and obtained a sinusoïdal change in the inductive voltage across the secondary coil on moving the magnets.
The field of the earth gave an inductive voltage with the frequency of
rotation of the apparatus (hence the eighth sub-harmonic), notwithstanding the rotational symmetry of the apparatus. This subharmo-
nic, however, could be decreased. Firm fastening of the screws, which
keep the system together, appeared to be necessary in connection
with vibrations. The apparatus has been placed upon wooden sup-
ports and a wooden base in order to keep the quantity of metal, not
moving with the magnets, small.

Of the apparatus constructed this is the only one in which the
temperature can be easily varied. E.g. the transition to the ferromagnetic state of iron, cooling from red heat, could easily be observed. Also the effect of the paramagnetic ferric sulfate heated from
liquid air to room temperature could be shown to increase by a
factor 3. The apparatus seems to be suitable for measurements at
higher temperatures.

To avoid disturbing fields caused by the motor, we constructed and
used a turbine. A P e l t o n wheel with 8 shovels with a surface of
100 mm² and a distance to the axis of 50 mm is used; 4 nozzles, with
a diameter of 5 mm, furnish us the waterjets. With a supply of 1 dm³
water per second the number of revolutions is 2200 per minute, the
power 0.1 kW.

Apparatus with moving sample [1])

As the quantity of substance necessary in the trial apparatus was

[1]) Mr. H. E. E r n s t cand. n.i., cooperated during the construction of this and the
preceding apparatus.

rather large, we tried to work with one single sample (fig. 8). We rotate a cylinder of which one side is filled with the sample around its axis in the pole gap of the magnet, also used in the trial apparatus. One of the pole pieces is divided into two parts, each carrying a coil.

Fig. 8

If the coils are oppositely connected, the curve drawn in fig. 4*f* represents the effects observed. Connecting the coils in the same direction, fig. 4*e* was found .The number of revolutions was chosen such that α became about 2 sec^{-1}. Applying (8) we find an effect of 0.3 mV. In reality the voltage is somewhat lower as the form of the curve is unfavourable.

In the apparatus the current fluctuations appeared to be reduced by a factor 100, which factor could be measured by exciting the electromagnet with a.c. In connection with the instrument described in IV we investigated the effects of the iron circuit in alternating fields. Although the field strength of course is high, still the disturbances due to eddy currents and ferromagnetic losses were so large that even the paramagnetic effect could not directly be observed.

4. Two special applications

Absolute measurements will be possible with an apparatus built according to the lines sketched above. In fig. 9 P is a primary coil, furnishing a strong field. The secondary coils S and S' can indicate differences in the field strength. As we want to measure absolutely, iron has to be avoided and also phase differences so that we work ballistically. When S' is removed, the substance Sa of cylindrical shape can be pushed into the field. The difference in the fieldstrength is $\varkappa \mu_0 H_p$, where H_p is the field strength due to the primary coil. This effect can be compensated with a coil C tightly wound upon the cylinder, hence moving with Sa, and carrying a current which can be branched off from the primary current. In the case of compensation, obtained when the ratio of the current in the primary and the compensation coil is α, we get on neglecting the difference in the surface

$$\varkappa = \frac{K_c}{K_p} \cdot \alpha,$$

where K_p and K_c are the field strengths when the coils carry a current of 1 A. Removing the sample and coil S, we can measure the ratio of the field constants by applying a.c. If the voltage across S' is zero, then

$$\frac{K_c}{K_p} = \alpha',$$

where α' is the ratio of the currents through the coils in this case, and hence as a result

$$\varkappa = \alpha\alpha'.$$

As we have cylindrical symmetry in the apparatus, we can eliminate inhomogeneities in the coils by working in different positions. If we assume a field strength of the primary coil of $B = 0.1$ Vsec/m² and a surface of the secondary coil S of 1 m², then $\Phi = 0.1$ Vsec. Using an EL6 tube in combination with a ballistic galvanometer, one has such a sentivity that 1 pro mille of the paramagnetic effect can be detected. Without EL6 we just observed the paramagnetic effect.

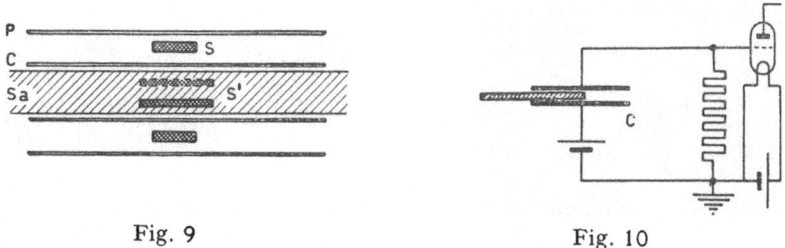

Fig. 9   Fig. 10

As another example we shall mention a *dielectric analogy*. Also here it is our intention only to point to a possibility rather than to give the result of substantial experimental work. In this generator a dielectric substance is periodically pushed in a condensor (fig. 10). The disk mentioned in 3, under the trial apparatus, was again used here, now as dielectric, while the capacity of the condensor was about 0.3 pF. Using some hundreds Hz, $1/\omega C = 10^9 \, \Omega$ and hence leakage is a great difficulty. Applying an electrometer triode and oscillograph as indicator we observed the effect of a solid substance.

In the more interesting case of a gas one can periodically change the density of the gas. As it is now possible to use a condensor of e.g. 500 pF, we get with $\varepsilon_r = 0.5 \cdot 10^{-3}$ a total variation of the capacity of 0.2^5 pF. Here we also just observed the effect. At the utmost it seems to be possible to use 1000 V across the condensor while probably 10^{-7} V can be measured. In the high frequency generator (II fig. 16) mostly used for these purposes a relative precision of 10^{-6} or 10^{-7} can be obtained; in this last instrument just as in the inductance apparatus with varying current the electric quantity, the voltage, is varied.

Chapter VI

CALCULATION OF AND ERRORS IN THE SUSCEPTIBILITY

1. Measurement of the susceptibility

The *measurement* of the susceptibility requires three measurements, two of which have to be performed with a substance of known susceptibility. The difference of these last two measurements furnishes the proportionality constant of the apparatus, always employing the same volume, while the difference with the unknown substance gives the required susceptibility.

A rather large difference in the magnetic effect is found on measuring air $(\varkappa(20°) = -36.9)$ and water $(\varkappa(20°) = 903.1)$, the total difference being $\Delta\varkappa = 940.0$. If one measures N_2 or CO_2 $(\varkappa(20°) = 0.4$ and 0.1 respectively), one has to close the plugs of the cuvette with fat as it appears that otherwise during the measurement a large part of the N_2 or CO_2 is replaced by air. Since water gives disturbances on account of dielectric losses, the mutual inductance apparatus has been calibrated with the ethyl-cyclohexane sample, often measured during the series with the torsion balance, as secondary standard.

One of the *errors* in the measurement is caused by the thermal expansion of the liquids. Therefore one always measures at the same temperature, most often 20°C, at which temperature also the density is measured. The *correction* in the case of the torsion balance is for pentane +1.6, for hexadecane +0.9 and for water +0.2 pro mille per °C temperature difference, being the relative thermal expansion of these liquids. In the inductance apparatus, as long as the liquid does not evaporate, the correction for the expansion is only half as great as in the torsion balance since the sample is rather well enclosed by the coil. Using a cuvette small compared with the secondary coil the weight will determine the effect entirely and the measurement will be independent of the temperature.

Much progress is made by compensating the main effect with a second substance. In the torsion balance we use a substance with $x = 808$, having a thermal expansion of $+1.1$ pro mille per °C, thus reducing the thermal effect by a factor 3 and making the correction most often negligible. Only in the case of water and air a correction has to be applied. In the inductance apparatus bromine is used for compensation since on filling the available volume its susceptibility is large enough to compensate the diamagnetic effect (see IV.2). As evaporation of the substances occurs, we weigh before and after the measurement and thus also can check the density.

The magnetic zeropoint of the empty carrier used in the torsion balance was checked with an ethyl-cyclohexane sample ($x = 795$) at regular intervals during a series. The cleaning and filling is easier then when using water, the temperature correction is less, while the effect of dissolved O_2 is also smaller. The shellac by which the coil is fixed to the carrier has to be dried and heated, since it otherwise changes too much when cleaning the cuvette. The proportionality constant of the apparatus changed no more than 2 pro mille during a series of some weeks.

In the later experiments the error in the adjustment of the balance was 0.2 pro mille, adding only a small contribution to the error in the susceptibility (≈ 1 pro mille), which probably is due to displacements of the balance or ferromagnetic effects. In the mutual inductance apparatus entirely different errors will be prominent, since here the adjustment causes the main error, while the cuvette can be very easily cleaned. Hence on comparing both methods it must be possible to reduce the effect of systematical errors.

The precision of the magnetic measurement is about 1 pro mille, to which the error in the density (< 0.5 pro mille) has to be added. Impurities in the form of isomers have a small effect according to the additivity rule. Assuming a relative proportion of isomers of $3 \cdot 10^{-2}$ and differences in the x of $3 \cdot 10^{-2}$, the change is 1 pro mille at the utmost.

2. Impurities

Ferromagnetic impurities are grave. As experiments (IV.3) show, $0.4 \cdot 10^{-8}$ g iron give 1 pro mille of our effect. In 8 g substance, which is the order of magnitude of the amount used, this is a relative quantity of $0.5 \cdot 10^{-9}$.

Measuring with the torsion balance, we observed that just after having started the measurements the deflection changed somewhat, indicating an increased diamagnetism of the substance. Iron can cause this effect; applying the field in the G o u y method, the particles can move towards the centre of the field. Here the force is zero so that the measurement no longer is affected. In the case of an air filling adhesion to the walls will prevent the motion of the particles. Indeed, two different series taught that the reproducibility of liquid measurements is better than that of the air points.

In order to investigate this point further, the following experiment was made. Filling a porcelan dish with water and some iron dust and filings, the particles sufficiently covered by water can move in the field and get sufficient velocity to shoot through the surface of the liquid. If small drops surround the particles, these can only direct themselves.

To prevent these effects the cuvette is filled through a funnel in a magnetic field. By doing so the effects disappeared and the results of the inductance apparatus agree with those of the balance.

We shall examine the effect of this filter, assuming that particles are present in the liquid which can give each 1 pro mille of our effect. When the forces in the liquid due to viscosity push the particles along the magnet, also a magnetic force is exerted upon them. As will be proved, this magnetic force is some orders of magnitude larger than the mechanical force.

Filling the cuvette with a volume of $4 \cdot 10^3$ mm³ through a funnel with a stem cross-section of 7 mm² in about 40 sec, the velocity of the liquid is $v = 14$ mm/sec. If the particles do not pass the filter, the mechanical force exerted upon them is

$$F = 6\pi\eta\, rv = 1.3 \cdot 10^{-9} \text{ N},$$

substituting $r = 5\mu$, r being the linear dimension of the particles (see IV.3g) and η the coefficient of viscosity. The magnetic force is

$$F = mJ \frac{dH}{dx} = 0.4 \cdot 10^{-5} \text{ N},$$

assuming $J = 0.5$ Vsec/m², while $m = 0.4 \cdot 10^{-11}$ kg, $H = 0.4 \cdot 10^5$ A/m (500 G). As is obvious for these particles, we have enough security as concerns the effect of this filter.

As the substances described in VIII.3 have been treated with

6

Ni-katalyser, a special test has been carried out with a substance not purified from Ni.

A general study of literature already affirms the existence of this error due to ferromagnetic impurities. While the values found for liquids never differ more than 0.1 of the effect, solid diamagnetic substances even have been found to be paramagnetic.

The occurrence of the paramagnetic *oxygen* causes a rather large dependence of x on the barometric pressure, x_{air} being —50 pro mille of our effect. A change of 15 mm in the pressure will give rise to 1 pro mille change so that we have to correct for this factor. The calibration point of water can be wrong if the water has not been boiled just before the experiment is performed. A quantity of oxygen can dissolve so that the diamagnetic effect is modified at most by 3 pro mille. The hydrocarbons, some of which were also boiled, showed no changes larger than 1 pro mille in this respect.

3. Comparison of the data

In order to suppress the occurrence of systematical errors, we measured with two apparatus. With the balance two series (1944 and 1946) have been performed. In the latter series we compensated the main magnetic effect and took care of displacements of the cuvette in connection with differences in the densities of the liquids (see III.3). Also some new samples were used. In graph 1 the difference

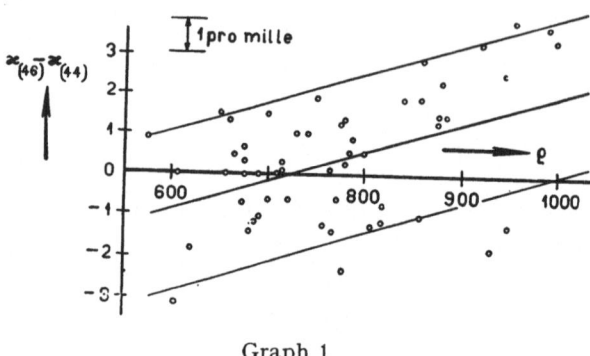

Graph 1

of x (46) and x (44) is plotted against ρ as far as the substances were measured twice. In graph 2 the averages over density intervals of 100 have been plotted, showing that probably the measurements of

1944 have an inaccuracy between.—1 and $+2\frac{1}{2}$ pro mille for extreme densities. We corrected the values of 1944 according to this graph and averaged, giving the values of 1946 twice the weight. In VIII graph 2 for the ethylalcohol—water mixture both values have been

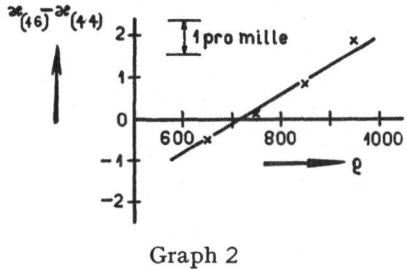

Graph 2

used, now agreeing within the precision of measurement. As graph 1 shows, the main part of the differences of the corrected values is less than $2\frac{1}{2}$ pro mille, $1\frac{1}{2}$ pro mille for x (44) and 1 pro mille for $x(46)$ (viz. the points between the two straight lines). In table I, II, III and IV the results for the pure compounds are given.

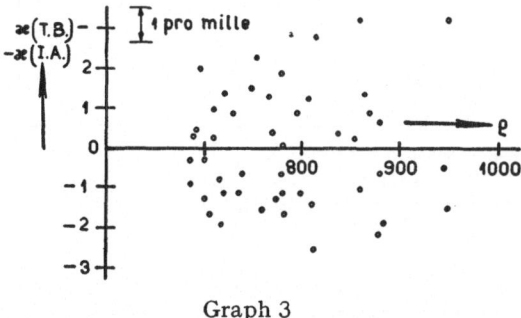

Graph 3

Graph 3 shows that between the difference of the measurements of the balance (T.B.) and the inductance apparatus (I.A.), and the density no correlation exists. Most compounds show no larger difference than 3 pro mille. It has to be noticed, however, that a hydrocarbon instead of water was used for the calibration of the inductance apparatus. The average value of x applied in the calculation of χ will have a precision of the order of one pro mille.

TABLE I [1])

	$\varkappa$			χ	χ_M		
	T.B.	I.A.	average		meas.	calc.	meas.-calc.
n-pentane	687.6		687.6	1.0981	79.23	78.6	+0.6
2-methyl-butane	695.0		695.0	1.1215	80.92	79.9	+1.0
n-hexane	712.1		712.1	1.0799	93.05	92.9	+0.2
2-methyl-pentane . . .	716.9		716.9	1.0975	94.57	94.2	+0.4
3-methyl-pentane . . .	731.7		731.7	1.1013	94.90	94.2	+0.7
2, 2-dimethyl-butane . .	721.8		721.8	1.1118	95.80	95.5	+0.3
2, 3-dimethyl-butane . .	735.5		735.5	1.1115	95.78	95.5	+0.3
n-heptane	730.9	731.0	731.0	1.0690	107.11	107.2	—0.1
2-methyl-hexane	734.0		734.0	1.0815	108.37	108.5	—0.1
2, 2-dimethyl-pentane . .	735 0		735.0	1.0907	109.29	109.8	—0.5
2, 3-dimethyl-pentane . .	762.8		762.8	1.0974	109.96	109.8	+0.2
2, 4-dimethyl-pentane . .	738.4		738.4	1.0972	109.94	109.8	+0.2
2, 2, 3-trimethyl-butane .	764.8		764.8	1.1081	111.03	111.1	0.0
n-octane	747.5	747.7	747.6	1.0631	121.43	121.4	0.0
3-methyl-heptane . . .	760.2	761.7	761.0	1.0781	123.14	122.7	+0.4
2, 3-dimethyl-hexane . .	775.1	774.1	774.6	1.0867	124.12	124.0	+0.1
3, 4-dimethyl-hexane . .	784.7	783.4	784.1	1.0898	124.48	124.0	+0.5
2, 5-dimethyl-hexane . .	751.0	749.2	750.1	1.0798	123.33	124.0	—0.7
2, 2, 3-trimethyl-pentane	786.3	787.1	786.7	1.0987	125.49	125.3	+0.2
2, 2, 4-trimethyl-pentane	748.3	749.0	748.7	1.0819	123.57	125.3	—1.7
n-nonane	760.5	761.7	761.1	1.0594	135.87	135.7	+0.2
n-decanc.	771.1	770.2	770.7	1.0555	150.18	150.0	+0.2
n-hexadecane	806.4	807.4	806.9	1.0412	235.77	235.7	+0.1
2, 2, 4, 7, 9, 9-hexamethyl-decane	828.8	828.7	828.8	1.0628	240.66	243.5	—2.8

examples of formulae (see also VII table VI)

pentane	$CH_3CH_2 \quad CH_2 \quad CH_2CH_3$
2, 2, 3-trimethyl-pentane	$CH_3C(CH_3)_2CH(CH_3)CH_2CH_3$
n-hexadecane	$CH_3(CH_2)_{14}CH_3$

The density adds probably no more than 0.5 pro mille to this. The data given by W i b a u t (1939, reference in VIII) sometimes were used. The chemical error will be small for the highly purified compounds, viz. the aliphatic hydrocarbons, and not exceed one pro mille so that the error in χ_m will be of the order of 1.5 pro mille. The other substances may have a lower purity. Their density and refractive index were observed and compared with literature. The chemical error may be some pro mille, the precision of χ_M 2 to 3 pro mille.

---

[1]) The compounds mentioned in table I and II, were put at our disposal by Prof. dr. ir. H. I. W a t e r m a n and the Bataafsche Petroleum Maatschappij.

TABLE II

	$\varkappa$			χ	χ_M		
	T.B.	I.A.	average		meas.	calc.	meas.-calc.
cyclopentane	790.4		790.4	1.0604	74.37	72.7	+1.7
methyl-cyclopentane . .	785.0	784.5	784.8	1.0478	88.18	88.3	—0.1
1, 2 and 1,3-dimethyl-cyclopentanes	782.8	781.4	782.1	1.0406	102.17	103.9	—1.7
cyclohexane	768.1	768.7	768.4	0.9868	83.05	83.5	—0.4
methyl cyclohexane . .	775.8	777.6	776.7	1.0100	99.16	99.1	+0.1
ethyl-cyclohexane. . . .	794.7	794.7	794.7	1.0201	114.47	113.3	+1.1
isopropyl-cyclohexane. .	819.7	820.8	820.3	1.0218	128.99	128.9	+0.1
tert. butyl-cyclohexane .	837.5	838.9	838.2	1.0311	144.62	145.5	+0.1
heptyl-cyclohexane . . .	823.0	825.4	824.2	1.0158	185.22	184.7	+0.5
octyl-cyclohexane . . .	827.9	825.3	826.6	1.0117	198.66	199.0	—0.4
decalin	856.6	855.9	856.3	0.9699	134.08	133.8	+0.3
dicyclohexyl	864.8	866.5	865.7	0.9771	162.49	162.4	+0.1
1,1-dicyclohexyl-nonane .	879.8		879.8	0.9965	291.50	292.2	—0.7
perhydro anthracene . .	902.0		902.0	0.9540	183.48	184.1	—0.6
benzene	773.3	774.1	773.7	0.8807	68.80	67.5	+1.3
toluene	776.5		776.5	0.9017	83.07	83.1	0.0
ethyl-benzene	797.3	796.4	796.8	0.9138	97.01	97.3	—0.3
n-heptyl-benzene	820.3		820.3	0.9581	168.90	168.7	+0.2
o-xylene.	809.0	809.6	809.3	0.9207	97.74	98.6	—0.9
m-xylene	784.9	782.0	783.5	0.9063	96.21	98.6	—2.4
p-xylene	781.8	782.8	782.3	0.9088	96.48	98.6	—2.2
1, 3, 5-trimethyl-benzene	795.8		795.8	0.9653	116.01	114.2	+1.8
1, 6-diphenyl-hexane . .	865.5	862.9	864.2	0.9058	215.90	216.0	—0.1
1, 1-diphenyl-nonane . .	870.7	872.1	871.4	0.9245	259.26	260.3	—1.0

saturated cyclic compounds		*aromatic compounds*	
cyclohexane	C_6H_{12}	benzene	C_6H_6
tert. butyl-cyclohexane	C_6H_{11} $C(CH_3)_3$	toluene	$C_6H_5CH_3$
decalin	$C_6H_{10} : C_4H_8$	o-xylene	$C_6H_4(CH_3)_2$
dicyclohexyl	$C_6H_{11} \cdot C_6H_{11}$	1.6-diphenyl-hexane	$C_6H_5(CH_2)_6C_6H_5$.
perhydro anthracene	$C_6H_{10} : (CH_2)_2 : C_6H_{10}$		

Though many authors mention a precision as high as 1 pro mille, the *literature* shows larger differences. A comparison is furnished by the calculation of the CH_2-group, of which values are found, ranging from 14.2 up to 14.9. Our value is 14.28, other values are 14.43 (reference 7); 14.66 (9); 14.28 (22); 14.63 (3); 14.69 (20). Concerning the mixtures some observers frequently found large deviations from the additivity rule, but these deviations decreased on increasing the precision of measurement; as a rule they are much smaller than 10 pro mille.

TABLE III

		$\varkappa$	χ	χ_M		
		T.B.		meas.	calc.	meas.-calc.
chloroform	$CHCl_3$	930	0.624	74.5		
carbon tetrachloride	$C\,Cl_4$	868	0.544	83.7		
prim. propyl chloride	$C_2H_5CH_2Cl$	795	0.898	70.5	70.1	+0.4
prim. n-butyl chloride	$C_3H_7CH_2Cl$	807	0.911	84.3	84.4	—0.1
sec. n-butyl chloride	$C_2H_5CHClCH_3$	798	0.915	84.7	84.4	+0.3
bromine	Br_2	1319	0.422	67.4		
methylene bromide	CH_2Br_2	1175	0.471	81.8		
bromoform	$CH\,Br_3$	1191	0.411	103.8		
ethyl bromide	C_2H_5Br	903	0.631	68.7	67.4	+1.3
ethylene dibromide	$CH_2\,Br\,CH_2Br$	1150	0.527	99.0	99.0	0.0
methylene iodide	CH_2J_2	1453	0.437	117.0	117.1	—0.1
methyl alcohol	$H\;CH_2OH$	666	0.840	26.9	27.3	—0.4
ethyl alcohol	$C\,H_3\,CH_2OH$	723	0.915	42.2	42.2	0.0
prim. propyl alcohol	$C_2H_5\;CH_2OH$	760	0.945	56.8	56.5	+0.3
prim. n-butyl alcohol	$C_3H_7\,CH_2OH$	768	0.948	70.3	70.7	—0.4
prim. n-hexyl alcohol	$C_5H_{11}CH_2OH$	800	0.973	99.5	99.3	+0.2
formic acid	$H\;COOH$	662	0.543	25.0	25.3	—0.3
acetic acid	$C\,H_3\;COOH$	698	0.665	40.0	40.2	—0.2
propionic acid	$C_2H_5\;COOH$	731	0.736	54.6	54.5	+0.1
n-butyric acid	$C_3H_7\;COOH$	752	0.785	69.2	68.7	+0.5
n-heptoic acid	$C_6H_{13}COOH$	787	0.855	111.3	111.6	—0.3
ethyl formate	$H\;COOC_2H_5$	667	0.729	54.0	53.4	+0.6
methyl acetate	CH_3COOCH_3	675	0.723	53.6	54.0	—0.4
ethyl acetate	$CH_3COOC_2H_5$	696	0.771	68.0	68.2	—0.2
iso-amyl acetate	$CH_3COO(CH_2)_2CH(CH_3)_2$	753	0.863	112.3	112.4	—0.1

Just as well as the results for the CH_2-group, the susceptibilities found by several authors for a definite compound show large differences. It appears that a number of compounds have been examined by more than one investigator. In table V we have collected the values of χ for those compounds of which several values are given in literature, including the value of the author. As concerns the method, it appears that, excepting by the author, only force methods have been used, but of different types. The results being obtained by investigators working in different laboratories, no systematic chemical errors would be expected, apart from the hygroscopic behaviour of acetone and ethyl alcohol.

TABLE IV

		$\varkappa$	χ	χ_M		
		T.B.		meas.	calc.	meas.-calc.
glycol	CH_2OHCH_2OH	877	0.784	48.7	48.6	+0.1
hexamethylene glycol	$CH_2OH(CH_2)_4CH_2OH$		0.896	105.9	105.7	+0.2
glycerol	CH_2OH $CHOH$ CH_2OH	979	0.778	71.7	71.0	+0.7
erythritol [1])	$CH_2OH(CHOH)_2CH_2OH$		0.759	92.7	93.4	(—0.7)
adonitol	$CH_2OH(CHOH)_3CH_2OH$		0.754	114.7	115.8	(—1.1)
sorbitol	$CH_2OH(CHOH)_4CH_2OH$		0.744	135.5	138.2	(—2.7)
mannitol	the same		0.767	139.7	138.2	(+1.5)
dulcitol	the same		0.775	141.2	138.2	(+3.0)
arabinose [1])	$OCH(CHOH)_3CH_2OH$		0.718	107.7	107.0	(+0.7)
xylose	the same		0.710	106.6	107.0	(—0.4)
glucose	$OCH(CHOH)_4CH_2OH$		0.716	129.0	129.4	(—0.4)
mannose	the same		0.718	129.3	129.4	(—0.1)
galactose	the same		0.719	129.5	129.4	(+0.1)
fructose	$CH_2OHCO(CHOH)_3CH_2OH$		0.716	129.0		
rhamnose	$OCH(CHOH)_4CH_3$		0.760	124.6	123.0	(+1.6)
aceton	CH_3COCH_3	580	0.732	42.5		
diethyl ether	$C_2H_5OC_2H_5$	671	0.934	69.2		
diethyl malonate	$COOC_2H_5CH_2COOC_2H_5$	750	0.720	115.3	115.0	+0.3
allyl alcohol	$CH_2 : CHCH_2OH$	679	0.794	46.1	46.1	0.0
cyclohexanol	$C_6H_{11}OH$	872	0.920	92.2	91.6	+0.6
phenyl chloride	C_6H_5Cl	860	0.777	87.4	87.6	—0.2
benzyl chloride	$C_6H_5CH_2Cl$	890	0.806	102.1	103.2	—1.1
α-napthyl chloride	$C_6H_4 : C_4H_3Cl$	992	0.831	135.2	134.7	+0.5
phenyl bromide	C_6H_5Br	936	0.625	98.2	99.1	—0.9
α-napthyl bromide	$C_6H_4 : C_4H_3Br$	1053	0.704	145.7	146.2	—0.5
phenyl iodide	C_6H_5J	1038	0.567	115.6	115.6	0.0
nitrobenzene	$C_6H_5NO_2$	759	0.631	77.7		
o-cresol	$C_6H_4(OH)CH_3$	872	0.823	89.0	89.7	—0.7
m- and p-cresol	the same	871	0.835	90.3	89.7	+0.6
methyl salicylate (o)	$C_6H_4(OH)COOCH_3$	840	0.712	108.4	107.9	+0.5

These values, especially that for acetone, show a large spread; this one even seems to decrease in the course of the years. On the other hand for several compounds fairly reproducible values are found, e.g. benzene and carbon tetrachloride. The fixing now of an average value will be advantageous in several respects. *a*. One can get an objective measure of the agreement between various observers. *b*. The results given by one observer can be compared with the average values. *c*. The compounds mentioned in table VI can be used as secondary standards.

[1]) The compounds up to rhamnose were put at our disposal by dr. J. v a n A l p h e n.

TABLE V

X-values; as an exceptional case the unit —10^{-11} is used

References		chloroform	carbon tetrachloride	acetone	methyl alcohol	ethyl alcohol	n-propyl alcohol	n-butyl alcohol	formic acid	acetic acid	propionic acid	n-butyric acid	ethyl formate	methyl acetate	ethyl acetate	n-propyl acetate	benzene	toluene	nitrobenzene
1	1931	624	542	750		934						792					885*	915*	628
2	1932	621	543	748								792					881		644*
3		626										786							
4																			
5	1933	627	544	745	876*	941*	940	937*	545	671		794*	728		770	810	877*	895*	640*
6																			
7																			
8	1934			730	847	920	948	950		672*				717	770	807	885*		627
9	1935		542		841				544	668					768*		885	907	628
10	1936		540*	750	834	920									770		884		627
11																			
12																			
13	1937				834	902*	948	952	561*	660*	743*				770		882		
14	1939			719		920												901	
15	1940		550*																
16					847				538	671							877*		633
17	1941	647*		715		922	949*	963*				787		724*			878		
18		620*			829*	902*	929*	945		664	736	785*		719	770	811	883	902	
19																			
20	1943	624	544	732	840	915	945	948	543	665	737		729	723	771	809	881	902	631
21	1946	626	543	736	842	920	944	949	544	668		789	729	720	770		882	903	631
average deviation		5	1	21	8	14	4	4	5	4	4	3	1	3	1	2	4	4	4

On comparing the values no correlation seems to exist between the precision mentioned by the authors and the deviation from the average values. Therefore this precision was not taken into account in weighting the observations. In order to prevent one value disturbing the estimate of the average, the values (indicated with *), lying far away from a preliminary mean, got half the weight. The number of these values was in the utmost $\frac{1}{3}$ of the total number of values. In the third column in table VI the number of values of the main group with weight one is given. The largest deviation in this group from the average value is taken as a measure for the error (see last row of table V). The values of B h a t n a g a r (reference 22) differed so much, e.g. for nitrobenzene and other compounds, reproducing fairly well for other authors, that they were not taken into account.

TABLE VI

χ-values, unit — 10^{-8}				
water	0.905		5 pro mille	
2 pro mille			chloroform.	0.625	5
carbon tetrachloride . . .	0.543	5	propyl alcohol	0.944	4
ethyl acetate	0.770	4	butyl alcohol	0.949	4
benzene	0.882	8	acetic acid	0.668	5
			propionic acid	0.737	2
10 to 20 pro mille			n-butyric acid	0.789	4
			ethyl formate	0.729	2
acetone.	0.736	8	methyl acetate	0.720	3
methyl alcohol	0.842	6	propyl acetate	0.809	3
ethyl alcohol	0.920	6	toluene	0.903	4
formic acid	0.543	4	nitrobenzene	0.630	6

In table VI the average values of table V are collected, likewise rounded off at one or two pro mille. The values of chloroform, formic acid and nitrobenzene have slightly been changed as some strongly deviating values still seemed to have too pronounced an effect. Likewise the estimated precision was somewhat rounded off, distinguishing three groups, apart from the calibration substance water. The table directly shows, that only for 15 substances the susceptibilities are known with a precision less than 10 pro mille.

Therefore in table III and IV, containing the compounds only measured once by the author, viz. with the balance, and probably not having the high purity of those of table I and II, the values have been rounded off at one or two pro mille. The values given in this thesis differ up to 5 pro mille from the averages; the average deviation being less than 2.5 pro mille. In the mean our values lie somewhat less than 1.5 pro mille below the averages.

A n g u s and H i l l (reference 16 and 20) published values differirg up to 20 pro mille, viz. for the alcohols. Still it seems to be difficult to say whether the averages or their values are wrong as their work is recent and seems to be thorough.

As a result we can state that in order to find reproducible values e.g. for the CH_2-group, more certainty should exist as concerns the data of table VI. Accordirg to the investigations of A n g u s and H i l l (20) the mean value for the CH_2-group for a series agrees well for different series and hence the differences in literature should not be due to the use of different series of substances by the several authors. The fixation of an average, however, with an absolute precision of 0.01 or even much better, as is sometimes done in literature, seems to be meaningless, the values of literature differing 0.3, the individual components of short chain length 0.1 or more.

REFERENCES TO TABLE V

1. R a n g a n a d h a m, S. P. (1931). Ind. Jour. Phys. **6**, 421; M a t h u r, R. N. (1931). Ind. Jour. Phys. **6**, 207.
2. R a o, S. R. and G. S i v a r a m a k r i s h n a n (1931). Ind. Jour. Phys. **6**, 509; B h a t n a g a r, S. S. and R. N. M a t h u r (1931). Phil. Mag. **11**, 914.
3. F a r q u h a r s o n, J. (1932). Nature **129**, 25; F a r q u h a r s o n, J. and M. V. C. S a s t r i (1937). Trans. Far. Soc. **33**, 1472.
4. C a b r e r a, B. and A. M a d i n a v e i t i a (1932). An. Soc. Esp. Fis. Quim. **30**, 528.
5. K i d o, K. (1932). Sci. Rep. Tôhoku Univ. **21**, 385; **22**, 835.
6. B o e k e r, G. F. (1933). Phys. Rev. (2) **43**, 756.
7. C a b r e r a, B. and H. F a h l e n b r a c h (1933). Z. Phys. **85**, 568.
8. R a o, S. R. and P. S. V a r a d a c h a r i (1934). Proc. Ind. Ac. (A) **1**, 77.
9. W o o d b r i d g e, D. B. (1935). Phys. Rev. **48**, 672.
10. S a l c e a n u, C. and D. G h e o r g h i u (1935). C. R. **200**, 120.
11. K i d o, K. (1936). Sci. Rep. Tôhoku Univ. **24**, 701.
12. S e e l y, S. (1936). Phys. Rev. **49**, 812.
13. R a o, S. R. and S. S r i r a m a n (1937). Phil. Mag. **24**, 1625; N e v g i (1938). J. Univ. Bombay **7**, 74.
14. R a o, S. R. and A. S. N a r a y a n a s i v a m y (1939). Proc. Ind. Ac. Sc. (A) **9**, 33.
15. M e e k s, W. W. and N. C. J a m e s o n (1940). Phys. Rev. (2) **57**, 71.
16. A n g u s, W. R. and W. K. H i l l (1940). Trans. Far. Soc. **36**, 923.
17. R a u t e n f e l d, F. v o n and E. S t e u e r (1941). Z. phys. Chem. (B) **51**, 39.
18. C a b r e r a, B. and H. C o l s o n (1941). C. R. **213**, 108.
19. C a b r e r a, B. (1941). Journ. chim. phys. **38**, 1.
20. A n g u s, W. R. and W. K. H i l l (1943). Trans. Far. Soc. **39**, 187.
21. B r o e r s m a, S. (1946). This thesis, table I, II, III and IV.
22. B h a t n a g a r and collaborators (1934). Z. Phys. **89**, 506; Phil. Mag. **18**, 449

CHAPTER VII

PURE COMPOUNDS

1. Homologous series

According to the *additivity rule* (I.5) the following holds:

$$\chi_M = \sum_k n_k \chi_M^k,$$

where n_k is the number of contributing elements of the kind k present in the molecule. In this chapter we shall try to derive the values χ_M^k, which are due to the several elements, from the experimental values of χ_M, obtained for the several compounds.

The main contribution is given by the CH_2-group. Its effect can be examined in the homologous series. In graph 1 we have plotted $\chi_M - n \cdot 14.3$ against n, the number of CH_2-groups present in the molecule. In a graphic representation the deviations are easily recognisable, the more so as we subtract the average effect of the CH_2-group (14.3) and hence obtain almost horizontal lines.

The normal aliphatic compounds (*a*) show but small deviations from a straight line. The mean slope corresponds to a value for the CH_2-group of 14.28, with a precision of about 2 pro mille. Only *n*-pentane shows an important deviation, 10^{-2} of the total value. The difference with hexane, due tot a CH_2-group, is 40 pro mille smaller than the average value for a CH_2-group in a long chain. The other lines, representing the average increment in other series of compounds, were drawn parallel to this line (apart from k).

On comparing compounds with an equal number of ramifications (*b*), larger deviations appéar. On the basis of their magnetic behaviour such compounds can be further divided into a few groups, viz. with respect to the place of the ramifications. We shall discuss this further in 3.

The series of cyclohexane (*c*) also show deviations as regards additi-

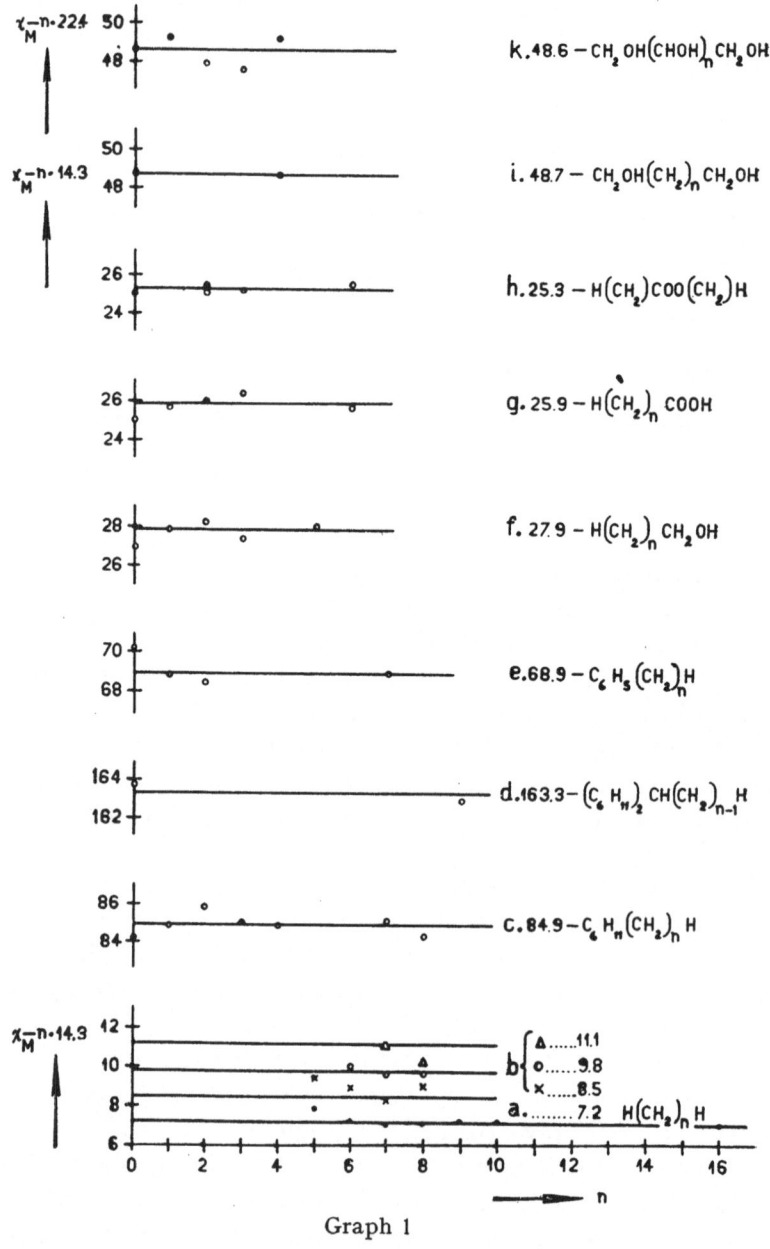

Graph 1

vity. For the higher compounds this may be due to impurities. For cyclohexane itself it can be real. In (d) compounds with 2 cyclohexane rings are represented. Benzene and its homologues (e) show a similar additivity. The values of the alcohols (*f*) and the acids (*g*) fit the drawn lines rather well. Line *h* represents the positions of the values of some esters. Glycol and homologous compounds (*i*) show good agreement with the assumed average effect of a CH_2-group.

Though lines with different inclination could also have been drawn, a change by e.g. 5 pro mille would show larger deviations from the points indicated. This last number has to be compared with the differences of the order of 30 pro mille of our value for the CH_2-group and values given in literature.

Line *k* gives the results for the alcohols from glycerol up to mannitol and isomers (the average of the isomers has been used). We subtract 22.4 for each CHOH-group. The deviations are not very large.

2. Calculation of several groups

For calculations of other compounds the estimate of a mean value for a ramification is very convenient. It appears to be about 1.3. For two ramifications roughly double this value is found, also on attaching the ramifications to the same carbon atom. Sometimes this value was already applied in the data of graph 1.

For a further analysis we distinguish primary, secondary, tertiary and quarternary C-atoms, according to the number of C–C-bonds starting from them. We shall use the value for a secondary C-atom in a long chain as a basis and give for the other sort of C-atoms the difference with this value. The introduction of a ramification then means that a secondary C-atom of the chain becomes a tertiary C-atom, while the introduced ramification contains, apart from secondary C-atoms, a primary C-atom at its end.

In fig. 1 the points of intersection of the several lines with the axis of ordinates directly yield the value for different groups. They are given in table I. It has to be noticed, that these values include both the effect of a group and a term connected with the structure.

The alipathic hydrocarbons yield the effect of two hydrogen atoms and two primary carbon atoms. The effect following from the cyclohexane series includes the effect of a ramification, just as well as that of the benzene homologues. The acid and alcohol series also give the effect of primary C-atoms.

TABLE I

H + primary C-atom	3.6	HCH$_2$OH + primary C-atom	27.9
ramification	1.3	HCOOH + primary C-atom	25.9
C$_6$H$_{12}$ + ramification	84.9	HCOOH + prim. C-atoms (esters)	25.3
C$_6$H$_{11}$ · C$_6$H$_{11}$ + ramific.	163.3	CH$_2$OH (CH$_2$-homologues)	24.3^5
C$_6$H$_6$ + ramification	68.9	CH$_2$OH (CHOH-homologues)	24.3
		CHOH	22.4

Examining formic acid and methyl alcohol (f, g, h) there is an indication that their points lie somewhat below the drawn lines. In acetic acid there is a primary C-atom and the COOH-group. In formic acid only the COOH-group is present, now connected to hydrogen. The methyl alcohol series is an analogous case. The average of the differences of the first compound in the series f, g and h and the drawn lines, 0.6 in the mean, we shall assume to be the effect of a primary C-atom. A portion of the effect, however, might be due to the change of the COOH-group. By doing so the other values χ_M^k kan be derived. In 4 the results for the atoms H, C and Br, thus obtained, will be compared with the results obtained for the pure elements, viz. H$_2$, Br$_2$ and solid carbon. If we should assume the effect of a primary C-atom to be zero, the value especially of hydrogen would become rather high.

From the average value 1.3 for a ramification there follows the effect of a tertiary C-atom, viz. 0.7. Two ramifications attached to the same C-atom give roughly twice the effect, viz. 2.6, and hence a quarternary C-atom is equivalent to two tertiary C-atoms.

As the primary C-atom contributes +0.6, it follows that a hydrogen atom, substituted in the molecule, adds +3.0. From the value 14.28 for CH$_2$ we then have left for carbon 8.28, for CH 11.28.

Now we can examine the saturated rings. The extra contribution due to the 6-ring in the cyclohexane series (table I) becomes 84.9 — — 6 CH$_2$ — ramif. = — 2.1. From dicyclohexyl (table I) follows 2 · — 2.3; decalin gives 2 · —2.1 and perhydro anthracene 3 · —2.4. This gives an average value of —2.2 for the 6-ring.

The 5-ring shows more irregular contributions to the effect of the available compounds viz. +3.0 for cyclopentane; +1.2 for methyl-cyclopentane, —0.4 for dimethyl-cyclopentane. Here the effect rather depends upon the compound under examination; as an order of magnitude we shall use +1.3.

Observing the homologues of benzene, we find an effect due to the ring of —0.2. For 1,6-diphenyl-hexane we obtain $2 \cdot$ —0.4, for 1,1-diphenyl-nonane $2 \cdot$ —0.8. Benzene itself and 1, 3, 5-trimethyl-benzene give positive values. The xylenes show the large influence of the positions of the ramifications. As in the case of cyclopentane, we find for the compounds, containing several ramifications attached to the ring, no constant value on calculating the effect of the ring.

The compounds containing halogens show an effect for the α-naphtyl-group, apart from the contributions of the atoms, of $+7.8$. We shall assume this value in the following calculation of the effect of Cl (see table II).

TABLE II

	$\chi_M - \chi_{Cl}$	χ_M	χ_{Cl}
$CHCl_3$	11.3	74.5	$3 \cdot 21.1$
CCl_4	8.3	83.7	$4 \cdot 18.8$
$C_2H_5CH_2Cl$	47.0	70.5	23.5
$C_3H_7CH_2Cl$	61.3	84.3	23.0
$C_2H_5CHClCH_3$	61.3	84.7	23.4
C_6H_5Cl	64.5	87.4	22.9
$C_6H_5CH_2Cl$	80.1	102.1	22.0
$C_6H_4 : C_4H_3Cl$	111.5	135.2	23.7

The first column contains the effect of the several contributions apart from Cl. The last column gives a mean value of 23.1 for the mono-substituents of Cl. The values for chloroform and carbon tetrachloride show that the Cl-atoms are rather deformed, compared with the other substances.

For bromine we find the values of table III.

TABLE III

	$\chi_M - \chi_{Cl}$	χ_M	χ_{Br}
Br_2		67.4	$2 \cdot 33.7$
CH_2Br_2	14.9	81.8	33.5
$CHBr_3$	11.9	103.8	30.6
C_2H_5Br	32.8	68.7	35.9
CH_2BrCH_2Br	29.8	99.0	34.6
C_6H_5Br	64.5	98.2	33.7
$C_6H_4 : C_4H_3Br$	111.5	145.7	34.2

The last four values give 34.6 as an average. From CH_2J_2 there follows an effect of $2 \cdot 51.1$ for J, from C_6H_5J 51.0.

The value for the COOH-group is found, considering the homologues of formic acid and the esters. The former give 22.3, subtracting from the value in table I the effect of H and the primary C-atom. The latter give 21.2; some esters contain 2, some 1 primary C-atom. We shall assume for COOH 22.3 and subtract 1.1 for esters. This would mean that the effect of H in CH_2 is different from that in the COOH-group which, however, could be expected.

Examining the alcohols, we find for the CH_2OH-group 27.9 — 3.6 = 24.3; from glycerol-homologues, with CH_2 substituted, there follows 24.3^5; with CHOH-groups substituted 24.3. This is in the mean 24.3. The CHOH-group gives an effect of 22.4. The calculated value for cyclohexanol then is 91.6, the experimental being 92.2. The only compound at our disposal having a double bond is allyl alcohol. The double bond adds —4.4 in this molecule.

Arabinose and xylose give 15.6 for the OCH-group, glucose, mannose and galactose +15.4 in the mean. Fructose gives 13.2 for the keton-group = CO. The investigations on the sugars, measured in solutions, suggest that the effect of stereo-isomerism probably will not exceed 10 pro mille, being about the precision of measurement here. The ring structure of the sugars was not taken into account. The effect of the CO-group in aceton is 6.7, which rather differs from that in fructose. Further investigation here might be profitable.

For the OH in the cresols there follows 9.6 on the average; applying this value, the effect of methyl salicylate fits fairly well.

3. Branched hydrocarbons

VI Table I shows the strong influence of the relative positions of the ramifications upon the susceptibility. As the results seem to be reliable, we shall introduce for a further analysis the distance of the primary and also of the tertiary C-atoms. As only ramifications containing one C-atom are present, the introduction of the relative position of primary and tertiary C-atoms is superfluous. Of interest are the primary and tertiary C-atoms of a ramification and the primary C-atoms at the end of the chain.

In table IV we give the difference of the branched and the normal compounds in the same sequence as in VI table I. Now we shall try in the utmost case to replace these 17 values by 7 other values, viz. the effect of an isolated ramification (r), the extra effect on attaching two primary C-atoms to the same C-atom in a chain (p_0), on attaching them to two neighbouring C-atoms (p_1) and finally on attaching them to two atoms with one secondary in

TABLE IV

Number of occurring increments and results for the several compounds											
Compound	r	p_0	p_1	p_2	t_0	t_1	t_2	$\Delta\chi_M$ meas.	calc.	ΔD meas.	calc.
$c\overset{c}{c}cc$	1	1	2	—1				1.7	1.7	3.8	3.7
$c\overset{c}{c}ccc$	1	1		2				1.5	1.4	4.3	3.7
$cc\overset{c}{c}cc$	1		2	1				1.9	1.9	1.1	2.2
$c\overset{c}{\underset{c}{c}}cc$	2	3	3		1			2.8	2.8	11.0	9.9
$cccc\underset{cc}{}$	2	2	4			1		2.7	3.1	3.5	3.5
$c\overset{c}{c}cccc$	1	1						1.3	1.2	4.3	3.7
$c\overset{c}{\underset{c}{c}}ccc$	2	3		3	1			2.2	2.2	8.3	9.9
$c\overset{cc}{c}cc$	2	1	3	2		1		2.9	3.0	1.8	2.1
$c\overset{c}{c}c\overset{c}{c}c$	2	2		4			1	2.8	2.2	8.9	7.3
$c\overset{cc}{c}cc\underset{c}{}$	3	4	6		1	2		3.9	3.9	6.3	5.9
$cc\overset{c}{c}cccc$	1		1					1.7	1.5	1.6	2.2
$c\overset{cc}{c}ccc$	2	1	2	1		1		2.7	2.6	1.9	2.1
$cc\overset{cc}{c}cc$	2		3	2		1		3.1	3.0	—0.2	0.6
$c\overset{c}{c}cc\overset{c}{c}c$	2	2						1.9	2.4	7.3	7.3
$c\overset{cc}{c}cc\underset{c}{}$	3	3	4	3	1	2		4.1	3.6	5.1	4.5
$c\overset{c}{c}c\overset{c}{c}c\underset{c}{}$	3	4		6	1		2	2.2	2.5	12.6	13.5

between (p_2); furthermore the extra effect of a quarternary C-atom compared with two tertiary apart (t_0), of two neighbouring tertiary atoms (t_1) and of two tertiary C-atoms with one secondary in between (t_2).

In table V this has been indicated. In the first three rows (p) the position of the primary C-atoms is referred to, in the following three (t) the position of the tertiary C-atoms. For a definite compound we

7

have to count the number of times such relative positions occur. The result of this counting is given in table IV. The occurrence of a p_2-value in n-pentane has been taken into account.

TABLE V

	r	χ_M +1.2	D +2.2
c xxxxx c	p_0	—	+1.45
xxxxx cc	p_1	+0.3	—
c c xxxxx	p_2	+0.1	—
x xxcxx x	t_0	—0.5	+1.1
xx xxccxx	t_1	—0.5	—3.8
x x xxcxcxx	t_2	—0.6	—
c xxcxx c	$(r + p + t)_0$	1.9	7.0
cc xxccxx	$(r + p + t)_1$	2.2	0.4
c c xcxcx	$(r + p + t)_2$	1.9	4.4
c c xcxxcx	$(r + p + t)_3$	2.4	4.4

By successive approximation the unknown values can be estimated. On calculating the values it appears that it is not necessary to introduce effects of ramifications farther apart; furthermore that p_0 can be dropped in the case of χ. In table V the values of p and t are given, in table IV the calculated susceptibilities. We get a better agreement than by applying an average value for a ramification. Though the profit is not so large on introducing another 5 constants, still we can expect that the data of table V represent a part of the occurring effects in the susceptibility of the compounds. In the calculation of χ_M, as given in VI table I, these data were not applied.

We also performed this analysis with the data of the F a r a d a y effect, obtained by the author on measuring the same hydrocarbons (for reference see VIII). Though the effects are stronger here, the agreement is better (D in table IV), compared with the application of one value for a ramification only. Furthermore no more than 4 constants are necessary here.

In the last four rows of table V we give the effects of two ramifications placed in different positions in a long chain. In both magnetic data an oscillating effect may be seen.

4. Results

Table VI contains the results as concerns the several contributions to the susceptibility. In VI table I, II, III and IV they have been applied. As is shown by the last column, the calculated values fit the measured ones rather well. The differences most often are smaller than 8 pro mille, less than 5 pro mille on the average. Some compounds, of which the measured value was not taken into account in the previous calculation of the group values, have not been calculated. The differences for the sugars having a precision of 10 pro mille, have been placed between brackets.

TABLE VI

CH_2	14.28	primary C-atom	0.6		
H	3.0	tertiary C-atom	0.7	COOH	22.3
C	8.28	double bond	(—4.4)	esters	—1.1
Cl	23.1	saturated 5-ring	(+1.3)	CH_2OH	24.3
Br	34.6	saturated 6-ring	—2.2	CHOH	22.4
J	51.1	benzene ring	—0.2	OCH	15.5
		naphtene ring	+7.8		

The value for the CH_2-group reproduces well. The values for the several compounds would not fit with this value on changing it by more than 5 pro mille, in the case of the aliphatic hydrocarbons on changing it by more than 2 pro mille. Examining compounds with short chains, including the compound with 1 C-atom, as is often done following literature, one can find higher values if the extra effect of the primary C-atom (0.6) is not distinguished. If one considers compounds containing even up to 6 C-atoms, one can still expect an effect of $0.6/6 = + 0.1$ in the CH_2-group.

As concerns the contribution of H, being 3.0, we recall (I. 3 and 4) the theoretical effect of H-atoms, viz. 2.98 and half the experimental value of pure H_2 2.52.

For some sorts of pure carbon one has found: $\chi = 0.63$, hence $\chi_M = 7.6$. Graphite shows a more aromatic character, its susceptibility depends strongly upon the orientation and also on the temperature, according to literature. We found in the organic compounds 8.3.

7

To check the order of magnitude of this effect, we shall estimate roughly the radius of the orbits of the 4 valency electrons in the carbon atoms. In the case $\chi_M^{(3)}$ the diamagnetic effect alone has the value 8.3, we find $r = 76$ pm. In aliphatic chains the distance of the carbon atoms is $2 \cdot 77$ pm. This fits as concerns the order of magnitude, the high frequency term being neglected.

For pure bromine we found $2 \cdot 33.7$; in organic compounds 34.6.

Now we shall consider the *aromatic compounds* somewhat further. For benzene we experimentally found 68.8 while the calculation with table VI gives 67.5. The calculation of naphtalene, applying the value $+7.8$ for the ring system, yields 114.6; probably naphtalene itself is also somewhat higher, about 116.

There is a second mode of calculation with the data of table VI too, viz. by applying the values for the saturated six-rings, for the double bonds and tertiary C-atoms. We then obtain 52.3 and 81.8 for benzene and naphtalene respectively, which means that extra effects, not mentioned in table VI, must be present, viz. $+16.5$ and $+34.2$.

Next we shall recall the results of K r i s h n a n (I.4). He found a large anisotropy, investigating solid aromatic compounds. In a direction parallel to the plane of the ring he obtained 43 and 69 for benzene and naphtalene respectively. The average susceptibilities were found to be 68 and 117.

This last result agrees fairly well with our data, being as mentioned 68.8 and 116. Therefore it speaks for itself that the extra effects found by us for the rings, 16.5 and 34.2, have to be related with an effect occurring in the aromatic compounds, viz. the circulation of the electrons through the ring. In first instance this does not effect the susceptibility parallel to the plane of the ring. Therefore one might ask why these last data, viz. 43 and 69 found by K r i s h-n a n, are still lower than those calculated by us: 52.3 and 81.8. As concerns the order of magnitude this can be understood: the isotropic effect of the ring arises from hydrogen and from 3 carbon valency electrons, instead of 4 carbon electrons, like in the aliphatic compounds, upon which our calculations were based.

CHAPTER VIII

MIXTURES

1. Liquids containing electric dipoles

In connection with the investigation of unknown mixtures we examined the change of the susceptibility on mixing pure compounds. In this case the additivity rule

$$\chi_M = \sum_k n_k \chi_M^k \qquad (1)$$

can be applied, in which n_k is the fraction of the kth component so that $\sum n_k = 1$. The mean molecular weight is

$$\overline{M} = \sum_k n_k M_k, \qquad (2)$$

as M is also additive. Rewriting (1) on applying (2) we find

$$\chi = \frac{1}{\overline{M}} \sum_k n_k \chi_M^k = \sum_k \frac{n_k M_k}{\overline{M}} \chi^k = \sum_k m_k \chi^k, \qquad (3)$$

where m_k is the fraction of mass of the kth component ($\sum m_k = 1$) and χ^k the susceptibility per unit mass. This means that χ is linearly related with the fraction of mass. Furthermore it easily follows that

$$\varkappa = \sum_k v_k \varkappa^k \qquad (4)$$

in which v_k is the fraction of volume of the kth component, provided no contraction occurs.

Of these relations (3) is the most convenient, as we most often weigh the components before mixing and the mean molecular weight is not required.

To enlarge the interaction effects, we mixed liquids of which one or both components contain electric dipoles, viz. ethyl alcohol—benzene, carbon tetrachloride—nitrobenzene, water—acetic acid and water—ethyl alcohol.

The values of the susceptibility of the pure compounds can be found in VI table I, Il, IlI or IV. In graph 1 and 2 the results are given. In all graphs the points representing the two pure compounds have been connected with two straight lines at a distance $+1.5$ and -1.5 pro mille from the average. We assume a precision of x of 1 pro mille and of the density and the concentration together of 0.5 promille.

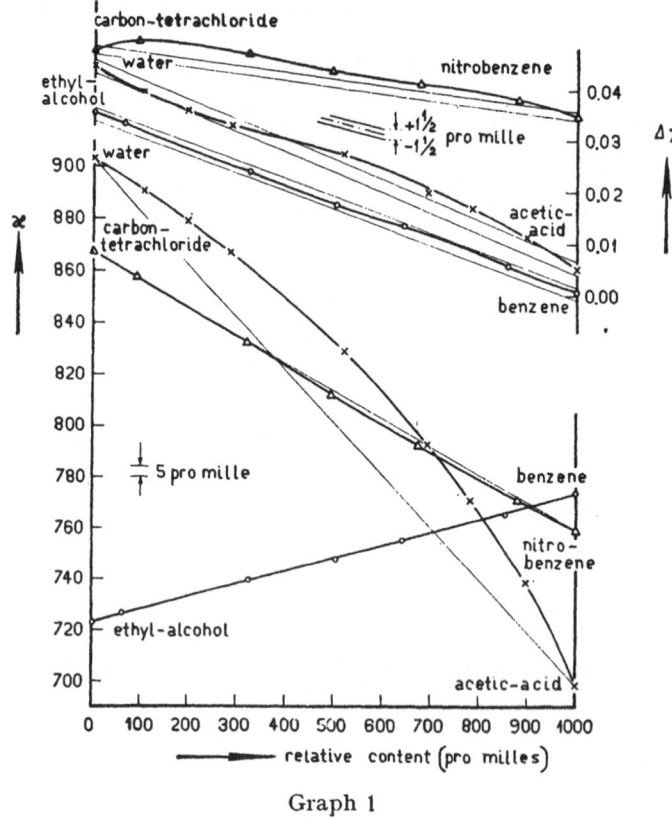

Graph 1

In graph 1 x and $\Delta\chi$, the difference of χ with a constant value for each of the systems, both have been plotted against the relative content of mass, as the latter only were known. We see that the deviation from a straight line of the values for ethyl acohol—benzene mixtures is less than 1 pro mille. The mixture of carbon tetrachloride and nitrobenzene shows a deviation in χ of 6 pro mille. Though the x-curve of water—acetic acid shows large departures from the straight line, still the χ-curve deviates no more than 4 pro mille.

Graph 2 shows the volume susceptibility of water-ethyl alcohol mixtures plotted against the relative content of mass and relative content of volume. There is a deviation of 25 and 40 pro mille respectively. It is noticeable that the two series of measurement, viz. 1944 and 1946, agree within the precision of measurement. They have not been indicated apart. The χ-curve, however, shows only

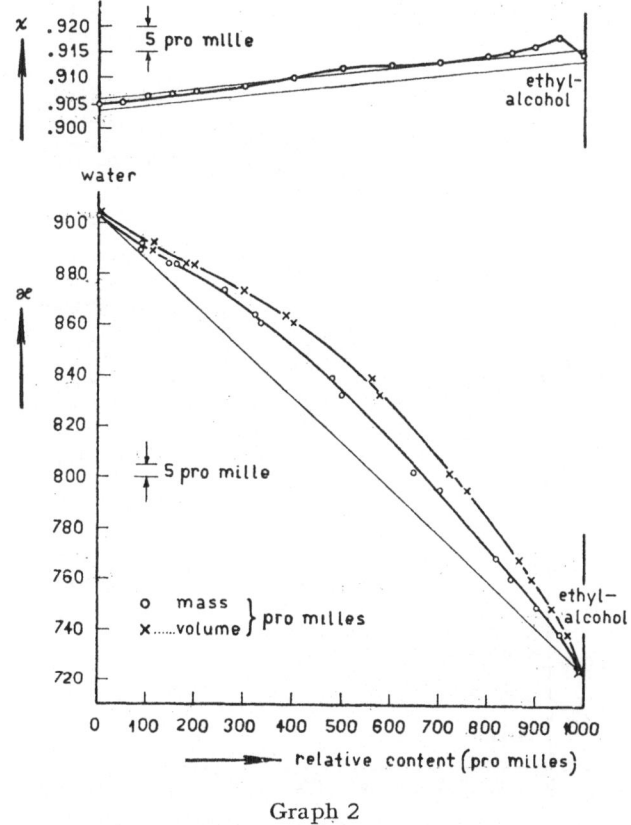

Graph 2

a deviation of 5 pro mille for high alcohol concentrations, for low concentrations the curve follows the straight line.

2. Hydrocarbons

We measured on 7 systems of mixtures and observe (graph 3) the following maximum deviations from the average of the values, obtained with the torsion balance (T.B.) and inductance apparatus (I.A.):

n-heptane—n-octane . . . . 0.5 pro mille,
n-heptane—2, 2, 4-trimethyl-pentane . . . . 1.0 ,,   ,,
n-heptane—n-hexadecane . . . . 1.0 ,,   ,,
n-heptane—methyl-cyclohexane . . . . 1.5 ,,   ,,
cyclchexane—methyl-cyclopentane. . . . . 1.0 ,,   ,,
cyclohexane—methyl-cyclohexane . . . . 1.5 ,,   ,,
decalin—methyl-cyclohexane . . . . 1.0 ,,   ,,

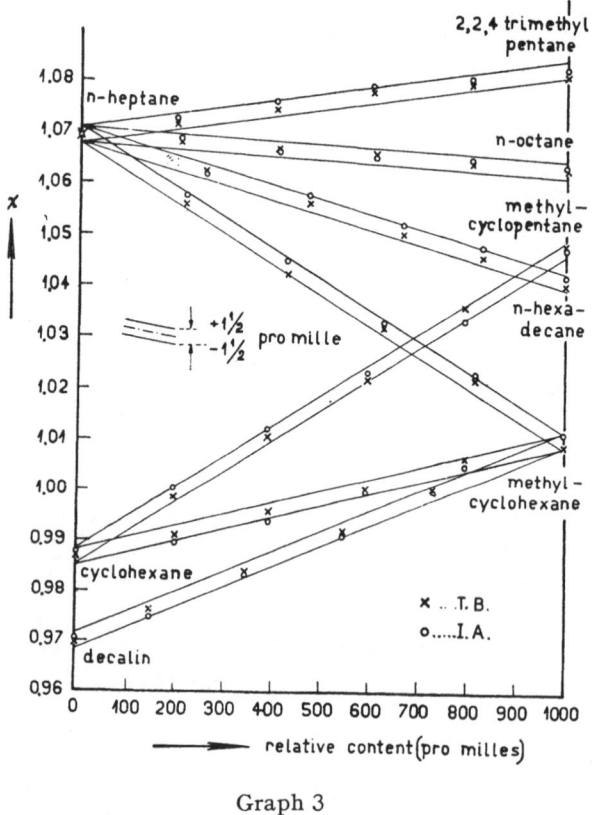

Graph 3

Hence we find linearity of the susceptibility with the concentration within the precision of measurement. This also occurs for the values of the two methods of measurement apart. As no dipoles are present and no contraction occurs, the deviations will probably be smaller.

We applied solutions of different concentrations for the measure-

ment of sugars and found by extrapolating to pure substances the
values given in VI table IV.

Furthermore we can expect that in the unkown hydrocarbon
mixtures the properties of the mixtures can be calculated by sum-
ming the contributions of the contributing elements.

3. Unknown mixtures

The following experiments deal with the research of those mixtures of hydrocarbons, as crude oil fractions, which cannot be further
seperated by means of the normal chemical or physico-chemical
seperation methods. The range of molecular weights in these mixtures is about 150 to 350 so that the number of C-atoms present will be
10 to 25, giving the possibility of an enormous number of isomers. It
is impossible to list them here and also unnecessary as their properties will differ only slightly. The differences are caused by special
groups such as rings, ramifications, double bonds, etc. Therefore it
will be better to mention these special features of the molecule, viz.
the fraction of carbon atoms forming rings, being tertiary or primary
atoms, etc. As the magnetic property of the mixture can be considered
to be calculable from the effects of the separate molecules, without
interaction effects, we have the possibility of treating these mixtures
statistically.

As is obvious, the susceptibility gives only one datum for the
mixtures of which more properties will be of interest. Fortunately
there are several physical constants of which the values are affected
by the molecular properties in a different manner. Furthermore by
means of slow chemical reactions, e.g. hydrogenation, some properties (double bonds) can be removed without changing the other
aspects of the constitution, e.g. saturated rings.

Therefore we only examine saturated hydrocarbons, consisting of
carbon and hydrogen atoms, rings and ramifications. The mean
molecular weight and molecular refraction are used, while in the
different fractions distilled from the same original ,,tin'' a certain
correspondence in sort of rings is assumed.

W a t e r m a n and his collaborators (1941) determined several
constants for three series: tin I (7 fractions), tin V (7 fractions) and
tin VIII (5 fractions). The density at 20°C was measured with a
precision of 0.3 pro mille. The mean molecular weight follows from

the decrease of the melting-point of benzene, furnishing an accuracy of about 10 pro mille.

As the precision of M is less than that of the other constants, we shall change the *expression of the additivity rule* somewhat by means of the equations (1) and (2).

$$\chi_M = n_C \chi_C + n_H \chi_H + \sum_k n_k \chi_k.$$

$$M = n_C M_C + n_H M_H.$$

Here $\sum_k n_k \chi_k$ only includes constitutive properties, not giving an addition in the molecular weight.

Furthermore in connection with the valency picture, the introduction of certain elements will change the number of H-atoms. From $C_n H_{2n+2}$ there follows $n_H = 2n_C + 2$; the introduction of double bonds, saturated rings and coupling of rings, will decrease n_H by 2 for each change so that

$$n_H = 2(n_C + 1 - n'_k), \tag{5}$$

in which n'_k indicates the number of these elements.

With these three equations n_H and n_C can be eliminated. We then obtain

$$(\chi - A)M = B + \sum_k n'_k (\chi'_k - B) + \sum_k n''_k \chi''_k, \tag{6}$$

where

$$A = \left(\frac{\chi}{M}\right)_{CH_2} \text{ and } B = \chi_{H_2} - M_{H_2} \cdot A. \tag{7}$$

A is the effect of the CH_2-group, the main element in long chains. B indicates the change in the content of hydrogen in the molecule. The property with a double accent changes the susceptibility, keeping the hydrogen content constant.

Formula (6) has been adapted to our problem. The influence of the moleculair weight agrees with its precision of measurement, while the main effect due to the CH_2-group is directly eliminated, its precise knowledge of course being necessary. As is obvious the formula holds for every additive property.

The measurement of the refractive index, at 20°C e.g. for the D-line, and the density, measured at the same temperature, yield us the *specific refraction*:

$$R_s = \frac{n^2 - 1}{n^2 + 2} \cdot \frac{1}{\rho}. \tag{8}$$

The molecular refraction $R = M R_s$ has been shown to be additive within a high precision, while only a slight number of contributions is of interest. The influence of the rings, extra effects of the primary and tertiary carbon atoms prove to be much smaller than the effect of the C and H atom itself (table I).

TABLE I

	R_D	$(R_F - R_C) \cdot 10^3$
CH$_2$	4.618	71
C	2.418	25
H	1.100	23
double bond	1.733	138
ramification	0.2 (+ or —)	8 ($\sim$)
ring	0.03 ($\sim$)	4 ($\sim$)

The effect of the double bond is large. The third column of table I, giving the difference of R for the F- and C-line, shows that here this effect is still stronger, compared with the other effects. The proper frequency related with the double bond lies more in the neighbourhood of the visible spectrum than the other proper frequencies. From the examination of $R_F - R_C$ it follows that the concentration of molecules with double bonds, present in our saturated mixtures, is of the order of 5 pro mille. If we only take one double bond per molecule into account, its effect is less than 1 pro mille.

Applying (8), only an A and B-value is of interest. If n_r is the number of rings, we find

$$(R_s - A_R) M = B_R (1 - n_r).$$

W a t e r m a n and collaborators carried out the analysis more empirically and most often used graphs. They could determine the *number of rings* with an accuracy of about 50 pro mille. As their published values had been rounded off, we employed the last formula and adapted the constants A and B; the formula then became

$$(R_s - 0.3294) M = - 1.50 (n_r - 1). \qquad (9)$$

The values of A and B thus found agree with the values found for pure compounds (see table I and eq. (7)), viz. $A = 0.3292$; $B = 1.54$. With (9) we calculated n_r anew, one decimal point further, and thus avoided jumps in the values of n_r, which could give difficulties in further calculations.

As is obvious, the application of this formula is easier than the use of graphs with the parameter n_r. In these graphs, in which R_s is plotted against M, the form of the lines with constant n_r resemble hyperbola (if $n_r \neq 1$). This agrees with (9); this equation also suggests that it is better to plot R_s against $1/M$.

Also the surface tension σ has been measured so that the *parachor*

$$P = \frac{\sigma^{1/4} M}{\rho_{liq} - \rho_{gas}}. \tag{10}$$

can be calculated, neglecting ρ_{gas}. This quantity is roughly additive according to S u g d e n and appears to be very sensitive for structural effects. Following W a t e r m a n and collaborators the *number of tertiary C-atoms* can be estimated by it, assuming a special sort of rings.

For the parachor we also performed a recalculation, giving

$$(P_s - 2.852) M = 26.6 - 29.0 (n_r + 0.1\, n_b), \tag{11}$$

when n_r indicates the number of rings, n_b the number of branches. These values agree with those, calculated from values given in literature for pure compounds. If we start with the data of W i b a u t and collaborators (1939), we find

$$(P_s - 2.850) M = 25 - 24\, n_r - 3\, n_t. \tag{12}$$

To derive the preceding equation, we have to introduce the special model with which the numbers n_b were determined by W a t e r-m a n (1941). This gives in connection with the coupling of rings

$$n_t = n_b + 2 (n_r - 1) + 1,$$

n_t being the number of tertiary C-atoms. By doing so we get

$$(P_s - 2.850) M = 28 - 30 (n_r + 0.1\, n_b),$$

which agrees with the first equation.

The *number of primary C-atoms* was calculated, using the same model, mentioned by L e e n d e r t s e (1938) (see also B r o e r s m a, W a t e r m a n, W e s t e r d ij k and W i e r s m a, 1943, page 124).

The values for n_r, n_t and n_p thus obtained are given in table II; n_t and n_p are less precise than n_r; the values have been rounded off in order not to exaggerate their precision.

Table II also contains the *susceptibilities* measured; χ has not been rounded off. The following step is the adaptation of (6) to these results. We then obtain 18 equations of the form

$$(\chi - A)\,M = B\,(1 - n_r) + \chi_r n_r + \chi_p n_p + \chi_t n_t, \qquad (13)$$

containing the variables A, B, χ_r, χ_p and χ_t. By successive approximation these values can be estimated.

TABLE II

	M	n_r	n_t	n_p	$\varkappa$	χ	$\dfrac{\chi_r}{\chi_b = 1.3}$	$\dfrac{a_r}{a_b = 2}$
Tin I								
1	166	0.9	2^5	2^5	832	1.0406	+0.2	—0.2
2	185	0.9	3	3^5	843	1.0428	0.0	—2.4
3	201	1.0	3	3	847	1.0390	+0.2	—2.3
4	235	1.0	3^5	3^5	851	1.0378	0.0	—2.3
5	247	1.1	4	4	857	1.0373	+0.1	—2.6
6	275	1.1	4	4	862	1.0359	+0.1	—2.6
7	304	1.3	5	4^5	869	1.0360	+0.4	—4.1
Tin V								
1	164	0.7	2	2^5	819	1.0402	—0.9	+0.3
2	175	0.8	2	2^5	832	1.0448	+1.2	+1.2
3	199	0.9	3	3	839	1.0407	+0.2	—4.8
4	219	0.9	3^5	3^5	847	1.0391	—0.1	—4.1
5	240	1.0	4	4	852	1.0365	—1.2	—3.4
6	277	1.1	4^5	4^5	852	1.0263	—2.9	—5.7
7	300	1.3	4^5	4	858	1.0222	—2.2	—4.8
Tin VIII								
1	176	1.1	3	2^5	849	1.0327	—0.3	—1.3
2	205	1.5	4	3	872	1.0326	0.0	—2.6
3	255	1.9	5	3^5	888	1.0280	+0.3	—2.5
4	303	2.2	7	5				
5	351	2.6	8^5	5	907	1.0219	—0.5	—2.1
effect 5-ring							(+1.3)	1.1
effect 6-ring							—2.2	—4.0
difference when changing χ_M or D by 1 pro mille							0.2	0.2

It appears that A agrees within 2 pro mille with the results for the pure substances and B within 5 pro mille, while the order of magnitude of χ_p and χ_t fits well. χ_r appears to lie between the value for a 5-ring and a 6-ring. This makes it probable that the values found in reality are the same as found for the pure substances and that deviations can be due to insufficient material. Especially tin V shows a large spread in the values, but the data of tin I and tin VIII seem to be fairly reliable.

Applying now the values found in VII, we can start with

$$(\chi - 1.018) M = 3.95 (1 - n_r) + \chi_r n_r + \tfrac{1}{2}(n_p + n_t) \chi_b + \varepsilon, \quad (14)$$

where $\chi_b = \chi_p + \chi_t$, while $\varepsilon = \tfrac{1}{2}(n_t - n_p)(\chi_t - \chi_p)$, small according to VII table VI, can be filled in.

Now only χ_r and χ_b are left. As both 5-rings and 6-rings occur in the mixtures, we only assume the ratio n_{5r} to n_{6r} to be constant for each of the tins apart. In the preliminary calculation we assumed it to be constant for all mixtures. We then examine the reproducibility of χ_r for the several mixtures of one tin, applying some different values for χ_b.

Observing the sum of the absolute deviations of χ_r from an average value for one tin, we find minima for $\chi_b = 1.3 \pm 0.1^5$, while then $\chi_r = + 0.1$ for tin I and $\chi_r = - 0.1$ for tin VIII. The last two values agree within the precision of the analysis. According to VII table VI $\chi_{6r} = - 2.2$ and an average value of $\chi_{5r} = + 1.3$, so that somewhat more 5-rings than 6-rings might be present. We wish to recall however, that the results are obtained by statistically examining the properties of the several fractions; their differences do not exceed 20 pro mille of the effect. In table II χ_r is given.

The author applied (6) also to the constant D_s, as given by the *Faraday effect* (see B r o e r s m a and collaborators, 1943). The results found here, as concerns the constants A and B of D_s, also agreed with the data obtained for pure compounds. The analysis further showed that the effect of the ring, a_r (last column in table II), as present in the mixtures, was the same for the fractions 1—6 of tin I, and 2, 3 and 5 of tin VIII. This result agrees with that obtained with the susceptibility.

It further appeared that a_r was somewhat nearer to a_{6r} than to a_{5r}. For the susceptibility we found that χ_r lies nearer to χ_{5r} than to χ_{6r}, but in connection with the precision this is not incompatible. Tin V gave in both calculations irregularities, still we again calculated χ_r. Now table II shows the typical result that there is some correlation between χ_r and a_r, which suggests that it is not one of the magnetic methods, which is unreliable in this respect, but probably other data.

In the analysis of D_s it appeared that the value for a ramification, as applied in the calculation of the mixtures, was 2.0 while for the pure compounds we assumed 3.0, as the average of values ranging from

2.7 for compounds with one ramification, to 3.4 for compounds with several ramifications. For χ_b we find 1.3, agreeing with the 1.3 assumed for the pure compounds. This value is the average of values from 1.5^5 to 1.0 for compounds containing one or several ramifications respectively. Therefore the results, as given by considering D_s, and χ, are more compatible if we compare the values found in the mixtures for D_s and χ, viz. 2.0 and 1.3 with the effect in compounds containing a single ramification or perhaps separate ramifications, viz. 2.7 and 1.5^5. The reliability of these facts, however, should not be over-estimated.

As a *result* we can state that the magnetic data seem to be able to give information as concerns the structure. The results obtained for the mixtures, by applying also data given by other physical constants, agree in several respects with those obtained from measurements on pure compounds. The number of ramifications, as given in table II, found by applying the parachor, seems to be somewhat too high; furthermore 5-rings and 6-rings are present, in tin I and tin VIII probably in a ratio between 2 : 1 and 1 : 2.

REFERENCES

Broersma, S., H. I. Waterman, J. B. Westerdijk and E. C. Wiersma (1943). Physica **10**, 97.
Leendertse, J. J. (1938). Thesis Delft, 143.
Waterman, H. I. and collaborators (1941). Chem. Wkbl. **38**, 91 and 106.
Wibaut, J. P. and collaborators (1939). Rec. Trav. Chim. **58**, 329.

LIST OF IMPORTANT QUANTITIES

The same symbols were used if no difficulties could arise from it.

B,H	magnetic field vectors	Φ,φ	magnetic flux
J	intensity of magnetisation	$\mathbf{I},I$	electric current
M,m	magnetic moment	K	field strength per A

V	electric potential	R	resistance
f	filling-factor	C	capacity
Z	impedance	L,M	self-, mutual inductance

μ_0	permeability of vacuum	$\varkappa$	susceptibility per m^3
μ_r	relative permeability	χ	susceptibility per kg
α	demagnetising factor	χ_M	susceptibility per kilomol.

S	surface	ρ	density
v	volume	n,N	number
m	mass	M	molecular weight

F,F	force	D	torsional constant, dielectric displacement, magneto-optic rotation
M	moment		
ω,ν	frequency		

v	velocity	J,P	angular momentum
r,γ	radius	g	Landé factor

SUMMARY

In this thesis we describe apparatus for the measurement of *magnetic susceptibilities* ($\varkappa = \mu - 1$). Three methods have been applied, viz. a torsion balance, an inductance apparatus operated with a.c. and an inductance apparatus in which the substance moves with respect to the magnet.

The *torsion balance* has been made very stable, yet keeping sufficient sensitivity to measure with an accuracy $\Delta \varkappa = 0.7 \cdot 10^{-8}$ in a time of about 25 minutes, filling etc. included. The effect of the density upon the adjustment in connection with the gravitational force was observed and avoided. Applying the G o u y method and compensating systems, liquids can be measured easily.

With the *inductance apparatus* with varying current we obtained the same precision, the time of measurement being about 8 minutes. The apparatus has become somewhat more extensive than the preceding one. A great number of disturbing effects in the coils were observed and reduced. Liquids could be easily measured, solids probably would give no difficulties.

In an *apparatus with varying coefficient of induction* we investigated for purposes of measurement the effect of a moving substance in a magnetostatic field. Here also a calculation of the effect is given. Up to now the precision was a factor 10 smaller than that obtained with the other apparatus.

The first two forms of apparatus were applied to the measurement of *organic substances* at room temperature. The results agree within the precision of measurement, being 1 pro mille of the diamagnetic effect. Furthermore a comparison with the data of literature was made.

We checked the *additivity rule*, by which the susceptibility of a compound is composed of values for the several constitutive groups. This seems to be satisfied within 5 pro mille on the average.

Furthermore *mixtures* were investigated. In the case of liquids, containing electric dipoles, deviations of the susceptibility from a linear relation with the concentration up to 6 pro mille were observed. For hydrocarbons linearity within the precision of measurement was confirmed. The analysis of unknown mixtures of hydrocarbons was pushed further, applying the magnetic data.

With the third apparatus no precise measurements were carried out. It is suitable for measurements at other temperatures than room temperature.

STELLINGEN

I

De verklaring van W o o s t e r, dat de bijdrage van een aromatisch koolstofatoom tot het isotrope deel van de susceptibiliteit $^3/_4$ is van dat van een aliphatisch koolstofatoom, tèngevolge van een speciale ligging van de atomen in het molecuul, is onjuist.

W o o s t e r, W. A. (1938), A textbook on crystal physics, blz. 115.

II

Bij het ontwerpen van een meetmethode is het steeds van belang na te gaan of de voordelen van de verschilmeting of nulmethode wel tot hun recht gekomen zijn. Men kan trachten te bereiken dat slechts hetgeen, waarin men geïnteresseerd is, direct afgelezen wordt.

III

Bij intensiteitsmetingen, b.v. polarimetrie, worden vaak draaiende sectoren toegepast om een wisselende lichtflux te maken. Dit kan eenvoudiger bereikt worden met gasontladingsbuizen, die op wisselspanning branden. De amplitude van de lichtfluctuaties is in beide gevallen vergelijkbaar, de frequentie wordt het dubbele van die der gebruikte wisselspanning.

IV

Het is gevaarlijk begrippen uit een bepaalde wetenscnap critiekloos in een andere te gaan gebruiken, in het bijzonder, wanneer zowel natuur- als cultuurwetenschappen in het geding zijn. Een bijzonder geval, dat dit bevestigt, geeft b.v.: H u i z i n g a, J. (1937). De wetenschap der geschiedenis, blz. 54.

V

De scherpe physische gedefinieerdheid van magnetische grootheden, evenals de eenvoudige samenhang met de eigenschappen der
afzonderlijke moleculen enerzijds en hun gevoeligheid voor de struc-
tuur anderzijds, maken deze grootheden zeer geschikt voor chemische analyse.

VI

In verband met de overschatting der precisie is voor de meeste
physische resultaten het aangeven van de waarschijnlijke fout niet
zeer doelmatig. Gebruik makende van de uitdrukking voor de fout in
de afzonderlijke metingen $F\sqrt{\Sigma v^2/(n-1)}$ liggen bij een Gaus-
sische verdeling bij de waarschijnlijke fout $(F = 0{,}67)$ 50% van de
meetresultaten binnen deze grenzen, voor $F = 2 : 95\%$. De fout
in het gemiddelde resultaat wordt $\sqrt{n}$ maal kleiner genomen, dus
$F\sqrt{\Sigma v^2/n(n-1)}$, hetgeen bij b.v. 10 metingen betekent dat resp.
17 en 47% der metingen binnen de grenzen vallen. Wij zouden
aan $F = 2$ de voorkeur willen geven.

VII

Bij nader onderzoek van enige literatuurgegevens blijkt dat de
verschillen in de resultaten van diverse methodes, m.a.w. de syste-
matische fouten van de afzonderlijke methodes in doorsnede meer
overeenstemmen met de toevallige fouten berekend met $F = 2$
dan met $F = 0{,}67$ (de waarschijnlijke fout).

VIII

De argumenten waarmede S u g d e n het bestaan van een ad-
ditiviteitsregel voor de parachoor toelicht, zijn niet zeer steek-
houdend.

S u g d e n, S., The parachor and valency,
blz. 30.

IX

.Ondanks het grote aantal studenten zal de vrijheid bij de studi
zo groot mogelijk dienen te blijven. Het aantal tentamina en ook d
contrôle bij practica beperke men tot een minimum.

X

Bij het hoger onderwijs zal men om aan de onderwerpen, die aan
de orde zijn, toe te kunnen komen zich tot de grote lijnen moeten
beperken, de afgesloten gedeelten overlatende aan andere onder-
wijsinstellingen. Met name kan dit gelden voor: chemische analyses,
het uitvoeren van meerdere proeven met eenzelfde strekking en
mogelijkerwijs ook voor het tekenen bij de meer constructieve af-
delingen, zodra dit routinewerk gaat worden.